Status and Understanding of Groundwater Quality in the Santa Clara River Valley, 2007: California GAMA Priority Basin Project

By Carmen A. Burton, Joseph Montrella, Matthew K. Landon, and Kenneth Belitz

A product of the California Groundwater Ambient Monitoring and Assessment (GAMA) Program

Prepared in cooperation with the California State Water Resources Control Board

Scientific Investigations Report 2011–5052

U.S. Department of the Interior
U.S. Geological Survey

U.S. Department of the Interior
KEN SALAZAR, Secretary

U.S. Geological Survey
Marcia K. McNutt, Director

U.S. Geological Survey, Reston, Virginia: 2011

For more information on the USGS—the Federal source for science about the Earth, its natural and living resources, natural hazards, and the environment, visit http://www.usgs.gov or call 1–888–ASK–USGS.

For an overview of USGS information products, including maps, imagery, and publications, visit http://www.usgs.gov/pubprod

To order this and other USGS information products, visit http://store.usgs.gov

Suggested citation:
Burton, C.A., Montrella, Joseph, Landon, M.K., and Belitz, Kenneth, 2011, Status and understanding of groundwater quality in the Santa Clara River Valley, 2007—California GAMA Priority Basin Project: U.S. Geological Survey Scientific Investigations Report 2011–5052, 86 p.

Contents

Contents—Continued

Figures

Figures

Tables

Conversion Factors, Datums, and Abbreviations and Acronyms

Conversion Factors

Inch/Pound to SI

Multiply	By	To obtain
	Length	
inch (in.)	2.54	centimeter (cm)
inch (in.)	25.4	millimeter (mm)
foot (ft)	0.3048	meter (m)
mile (mi)	1.609	kilometer (km)
	Area	
square mile (mi^2)	2.590	square kilometer (km^2)
	Flow rate	
gallon per minute (gal/min)	0.06309	liter per second (L/s)

SI to Inch/Pound

Multiply	By	To obtain
	Length	
millimeter (mm)	0.03937	inch (in.)
kilometer (km)	0.6214	mile (mi)
meter (m)	3.281	foot (ft)
	Volume	
liter (L)	0.2642	gallon (gal)
	Mass	
kilogram (kg)	2.205	pound avoirdupois (lb)

Temperature in degrees Celsius (°C) may be converted to degrees Fahrenheit (°F) as follows:

$$°F=(1.8×°C)+32.$$

Specific conductance is given in microsiemens per centimeter at 25 degrees Celsius (µS/cm at 25°C).

Concentrations of chemical constituents in water are given either in milligrams per liter (mg/L) or micrograms per liter (µg/L). Milligrams per liter is equivalent to parts per million (ppm) and micrograms per liter is equivalent to parts per billion (ppb).

cm^3 STP g^{-1}	cubic centimeters at standard temperature and pressure per gram
δE	delta notation; the ratio of the heavier isotope (i) to the more common lighter isotope of an element (E), relative to a standard reference material, expressed as per mil
mL	milliliter
pCi/L	picocuries per liter
per mil	parts per thousand
pmc	percent modern carbon
TU	tritium unit
%	percent

Conversion Factors, Datums, and Abbreviations and Acronyms—Continued

Datums

Vertical coordinate information is referenced to the North American Vertical Datum of 1988 (NAVD 88).

Horizontal coordinate information is referenced to the North American Datum of 1983 (NAD 83).

Abbreviations and Acronyms

AL-US	U.S. Environmental Protection Agency action level
BLS	below land surface
E	estimated or having a higher degree of uncertainty
GAMA	Groundwater Ambient Monitoring and Assessment Program
HAL-US	U.S. Environmental Protection Agency lifetime health advisory level
HBSL	health-based screening level
LRL	laboratory reporting level
LSD	land-surface datum
LT-MDL	long-term method detection level
MCL-CA	California Department of Public Health maximum contaminant level
MCL-US	U.S. Environmental Protection Agency maximum contaminant level
MDL	method detection limit
MRL	minimum reporting level
NL-CA	California Department of Public Health notification level
NWIS	National Water Information System (USGS)
PSW	public-supply wells
RPD	relative percentage difference
RSD	relative standard deviation
RSD5-US	U.S. Environmental Protection Agency risk-specific dose at a risk factor of 10^{-5}
SCRV	Santa Clara River Valley study unit
SCRVU	Santa Clara River Valley study unit understanding well
SMCL-CA	California Department of Public Health secondary maximum contaminant level
SMCL-US	U.S. Environmental Protection Agency secondary maximum contaminant level
TEAP	terminal electron acceptor process
US	United States
VSMOW	Vienna Standard Mean Ocean Water
>	greater than
≥	greater than or equal to
<	less than
≤	less than or equal to

Conversion Factors, Datums, and Abbreviations and Acronyms—Continued

Organizations

CDPH	California Department of Public Health (was California Department of Health Services prior to July 1, 2007)
CDWR	California Department of Water Resources
LLNL	Lawrence Livermore National Laboratory
NAWQA	National Water-Quality Assessment Program (USGS)
NWQL	National Water Quality Laboratory (USGS)
SWRCB	State Water Resources Control Board (California)
UWCD	United Water Conservation District
USEPA	U.S. Environmental Protection Agency
USGS	U.S. Geological Survey

Selected chemical names

Ammonia-N	ammonia as nitrogen
$CaCO_3$	calcium carbonate
DBCP	1,2-dibromo-3-chloropropane
DO	dissolved oxygen
DOC	dissolved organic carbon
EDB	1,2-dibromoethane (ethylene dibromide)
Fe	iron
Nitrate-N	nitrate as nitrogen
Nitrite-N	nitrite as nitrogen
Mn	manganese
PCE	perchloroethene (tetrachloroethene)
S	sulfur
TCE	trichloroethene
TDS	total dissolved solids
THM	trihalomethane
VOC	volatile organic compound

This page intentionally left blank.

Status and Understanding of Groundwater Quality in the Santa Clara River Valley, 2007: California GAMA Priority Basin Project

By Carmen A. Burton, Joseph Montrella, Matthew K. Landon, and Kenneth Belitz

Abstract

Groundwater quality in the approximately 460-square-mile Santa Clara River Valley study unit was investigated from April through June 2007 as part of the Priority Basin Project of the Groundwater Ambient Monitoring and Assessment (GAMA) Program. The GAMA Priority Basin Project is conducted by the U.S. Geological Survey (USGS) in collaboration with the California State Water Resources Control Board and the Lawrence Livermore National Laboratory. The Santa Clara River Valley study unit contains eight groundwater basins located in Ventura and Los Angeles Counties and is within the Transverse and selected Peninsular Ranges hydrogeologic province.

The Santa Clara River Valley study unit was designed to provide a spatially unbiased assessment of the quality of untreated (raw) groundwater in the primary aquifer system. The assessment is based on water-quality and ancillary data collected in 2007 by the USGS from 42 wells on a spatially distributed grid, and on water-quality data from the California Department of Public Health (CDPH) database. The primary aquifer system was defined as that part of the aquifer system corresponding to the perforation intervals of wells listed in the CDPH database for the Santa Clara River Valley study unit. The quality of groundwater in the primary aquifer system may differ from that in shallow or deep water-bearing zones; for example, shallow groundwater may be more vulnerable to surficial contamination. Eleven additional wells were sampled by the USGS to improve understanding of factors affecting water quality.

The *status assessment* of the quality of the groundwater used data from samples analyzed for anthropogenic constituents, such as volatile organic compounds (VOCs) and pesticides, as well as naturally occurring inorganic constituents, such as major ions and trace elements. The *status assessment* is intended to characterize the quality of untreated groundwater resources in the primary aquifers of the Santa Clara River Valley study unit, not the quality of treated drinking water delivered to consumers.

Relative-concentrations (sample concentration divided by health- or aesthetic-based benchmark concentration) were used for evaluating groundwater quality for those constituents that have Federal and (or) California benchmarks. A relative-concentration greater than 1.0 indicates a concentration greater than a benchmark. For organic and special interest constituents, relative-concentrations were classified as high (greater than 1.0); moderate (greater than 0.1 and less than or equal to 1.0); and low (less than or equal to 0.1). For inorganic constituents, relative-concentrations were classified as high (greater than 1.0); moderate (greater than 0.5 and less than or equal to 1.0); and low (less than or equal to 0.5).

Aquifer-scale proportion was used as the primary metric in the *status assessment* for evaluating regional-scale groundwater quality. High aquifer-scale proportion is defined as the areal percentage of the primary aquifer system with relative-concentrations greater than 1.0. Moderate and low aquifer-scale proportions are defined as the areal percentage of the primary aquifer system with moderate and low relative-concentrations, respectively. Two statistical approaches, grid-based and spatially weighted, were used to evaluate aquifer-scale proportions for individual constituents and classes of constituents. Grid-based and spatially weighted estimates were comparable in the Santa Clara River Valley study unit (within 90 percent confidence intervals).

The *status assessment* showed that inorganic constituents were more prevalent and relative-concentrations were higher than for organic constituents. For inorganic constituents with human-health benchmarks, relative-concentrations (of one or more constituents) were high in 21 percent of the primary aquifer system areally, moderate in 30 percent, and low or not detected in 49 percent. Inorganic constituents with human-health benchmarks with high aquifer-scale proportions were nitrate (15 percent of the primary aquifer system), gross alpha radioactivity (14 percent), vanadium (3.4 percent), boron (3.2 percent), and arsenic (2.3 percent). For inorganic constituents with aesthetic benchmarks, relative-concentrations (of one or more constituents) were high in 54 percent of the primary aquifer system, moderate in 41 percent, and low or not detected in 4 percent. The inorganic constituents with aesthetic benchmarks with high aquifer-scale proportions were total dissolved solids (35 percent), sulfate (22 percent), manganese (38 percent), and iron (22 percent).

In contrast, the results of the *status assessment* for organic constituents with human-health benchmarks showed that relative-concentrations were high in 0 percent (not detected above benchmarks) of the primary aquifer system, moderate in 8.4 percent, and low or not detected in 95 percent. Relative-concentrations of the special interest constituent, perchlorate, were moderate in 12 percent of the primary aquifer system and low or not detected in 88 percent. Relative-concentrations of two VOCs—carbon tetrachloride and trichloroethene (TCE)—were moderate in 2.4 percent of the primary aquifer system. One VOC—chloroform (water disinfection byproduct)—was detected in more than 10 percent of the primary aquifer system but at low relative-concentrations. Of the 88 VOCs and gasoline oxygenates analyzed, 71 were not detected. Pesticides were low or not detected in 100 percent of the primary aquifer system. Of the 118 pesticides and pesticide degradates analyzed, 13 were detected and 5 of those had human-health benchmarks. Two of these five pesticides—simazine and atrazine—were detected in more than 10 percent of the primary aquifer system.

The second component of this study, the *understanding assessment*, was to identify the natural and human factors that affect groundwater quality on the basis of the evaluation of land use, physical characteristics of the wells, and geochemical conditions of the aquifer. Results from these analyses are used to explain the occurrence and distribution of selected constituents in the primary aquifer system of the Santa Clara River Valley study unit.

The *understanding assessment* indicated that water quality varied spatially primarily in relation to depth, groundwater age, reduction-oxidation conditions, pH, and location in the regional groundwater flow system. High and moderate relative-concentrations of nitrate and low relative-concentrations of pesticides were correlated with shallow depths to top-of-perforation, and with high dissolved oxygen. Groundwater of modern and mixed ages had higher nitrate than pre-modern-age groundwater. Decreases in concentrations of total dissolved solids (TDS) and sulfate were correlated with increases in pH. This relationship probably indicates relations of these constituents with increasing depth across most of the Santa Clara River Valley study unit. Previous studies have indicated multiple sources of high concentrations of TDS and sulfate and multiple geochemical processes affecting these constituents in the Santa Clara River Valley study unit. Manganese and iron concentrations were highest in pre-modern-age groundwater at depth and in the downgradient area of the Santa Clara River Valley study unit (closest to the coastline), indicating the prevalence of reducing groundwater conditions in these aquifer zones.

Introduction

Groundwater composes almost one-half of the water used for public supply in California (Hutson and others, 2004). To assess the quality of ambient groundwater in aquifers used for drinking-water supply and to establish a baseline groundwater quality monitoring program, the California State Water Resources Control Board (SWRCB) in collaboration with the U.S. Geological Survey (USGS) and Lawrence Livermore National Laboratory (LLNL) implemented the Groundwater Ambient Monitoring and Assessment (GAMA) Program (California Environmental Protection Agency, 2010, website at http://www.waterboards.ca.gov/water_issues/programs/gama). The statewide GAMA program currently consists of three projects: (1) the GAMA Priority Basin Project, conducted by the USGS (U.S. Geological Survey, 2010, website at http://ca.water.usgs.gov/gama); (2) the GAMA Domestic Well Project, conducted by the SWRCB; and (3) the GAMA Special Studies, conducted by LLNL. On a statewide basis, the GAMA Priority Basin Project primarily focused on the deep portion of the groundwater resource (primary aquifer system) and the SWRCB Domestic Well Project generally focused on the shallow aquifer systems. The primary aquifer system may be at less risk of contamination than shallow wells, such as private domestic or monitoring wells, that are closer to surficial sources of contaminants. As a result, concentrations of contaminants such as VOCs and nitrate can be higher in shallower wells than deeper wells (Nolan and Hitt, 2006; Zogorski and others, 2006; Landon and others, 2010).

The SWRCB initiated the GAMA Program in 2000 in response to a legislative mandate (State of California, 1999, 2001a, Supplemental Report of the 1999 Budget Act 1999–00 Fiscal Year). The GAMA Priority Basins Project was initiated in response to the Groundwater Quality Monitoring Act of 2001 (State of California, 2001b, Sections 10780–10782.3 of the California Water Code, Assembly Bill 599) to assess and monitor the quality of groundwater in California. The GAMA Priority Basin Project is a comprehensive assessment of statewide groundwater quality designed to help better understand and identify risks to groundwater resources, and to increase the availability of information about groundwater quality to the public. For the GAMA Priority Basin Project, the USGS, in collaboration with the SWRCB, developed the monitoring plan to assess groundwater basins through direct and other statistically reliable sampling approaches (Belitz and others, 2003; California State Water Resources Control Board, 2003). Additional partners in the GAMA Priority Basin Project include the California Department of Public Health (CDPH), the California Department of Pesticide Regulation (CDPR), the California Department of Water Resources (CDWR), and local water agencies and well owners.

The range of hydrologic, geologic, and climatic conditions that exist in California must be considered in an assessment of groundwater quality. Belitz and others (2003) partitioned the State into 10 hydrogeologic provinces, each with distinctive hydrologic, geologic, and climatic characteristics (fig. 1). All these hydrogeologic provinces include groundwater basins designated by the CDWR (California Department of Water Resources, 2003). Groundwater basins generally consist of relatively permeable, unconsolidated deposits of alluvial or volcanic origin. Eighty percent of California's approximately 16,000 active and standby drinking-water wells listed in the statewide database maintained by the CDPH (hereinafter referred to as CDPH wells) are located in designated groundwater basins within these hydrogeologic provinces. Some groundwater basins, such as the Santa Clara River Valley basin, cover large areas and are further divided into groundwater subbasins by CDWR. Groundwater basins and subbasins were prioritized for sampling on the basis of the number of CDPH wells, with secondary consideration given to municipal groundwater use, agricultural pumping, the number of historical leaking underground fuel tanks, and registered pesticide applications (Belitz and others, 2003). Of the 472 basins and subbasins designated by the CDWR, 116 priority basins, containing approximately 95 percent of the CDPH wells located in basins, were selected for the project. These priority basins and additional areas outside defined groundwater basins were grouped into 35 study units.

The Santa Clara River Valley GAMA study unit (hereinafter referred to as SCRV) is located in the Transverse and selected Peninsular Ranges hydrogeologic province (fig. 1) (Belitz and others, 2003). SCRV consists of eight groundwater basins (Ventura River Valley, Ojai Valley, Upper Ojai Valley, Santa Clara River Valley, Pleasant Valley, Las Posas Valley, Arroyo Santa Rosa Valley, and Simi Valley) (fig. 2).

Purpose and Scope

This report is one in a series of published and planned Scientific Investigation Reports presenting the *status* and *understanding* of current water-quality conditions in GAMA Priority Basin Project study units. Tabulated USGS data are available from several study units and are available as USGS Data Series reports (for example, Montrella and Belitz, 2009), and planned subsequent reports will address changes or *trends* in water-quality across time.

The *status* and *understanding assessments* for the SCRV study unit are presented in this report. The purposes of this report are to provide a (1) study unit description: briefly describe the hydrogeologic setting of the SCRV study unit, (2) *status assessment:* assessment of the current status of untreated-groundwater quality in the primary aquifer system in the SCRV study unit, and (3) *understanding assessment:* identification of the natural and human factors affecting groundwater quality, and explanation of the relations between water quality and selected potential explanatory factors. An explanation of the causative factors of any relations between water quality and explanatory factors is beyond the scope of this report.

The *status assessment* in this report includes analysis of water-quality data from 42 wells selected for sampling by the USGS within spatially distributed grid cells across the SCRV (hereinafter referred to as USGS-grid wells). The USGS-grid wells mostly were public-supply wells (PSW) but included other wells with perforation intervals similar to wells listed in the CDPH database. Samples were collected from USGS-grid wells for analysis of anthropogenic constituents, such as volatile organic compounds (VOCs) and pesticides, as well as naturally occurring constituents, such as major ions, nutrients, and trace elements. Water-quality data from the CDPH database also were used to supplement data collected by USGS for the GAMA Program. The resulting set of water-quality data from USGS-grid and selected CDPH-grid wells were considered to be representative of the primary aquifer system in the SCRV study unit.

To provide context, the water-quality data discussed in this report were compared to California and Federal regulatory and non-regulatory benchmarks for treated drinking water. The assessments in this report are intended to characterize the quality of untreated groundwater resources in the primary aquifers in the study unit, not the treated drinking-water delivered to consumers by water purveyors. Benchmarks apply to treated water that is delivered to the consumer, not to groundwater.

The *understanding assessment* for SCRV includes data from 11 wells sampled by USGS for the purpose of understanding (hereinafter referred to as USGS-understanding wells). Some of the USGS-understanding wells selected for this purpose had perforations in shallow or deep zones, above or below the primary aquifer system. Potential explanatory factors examined included land use, well depth and depth to top-of-perforations, groundwater age, and geochemical-condition indicators (redox status). A comprehensive analysis of all possible explanatory factors is beyond the scope of this report.

Water-quality data for samples collected by the USGS for the GAMA Program in the SCRV study unit and details of sample collection, analysis, and quality-assurance procedures were reported by Montrella and Belitz (2009). Using the same data, this report describes methods used in designing the sampling network, identifying CDPH data for use in the status assessment, estimating aquifer-scale proportions, analyzing ancillary datasets, classifying groundwater age, and assessing the status and understanding of groundwater quality and its relation to selected explanatory factors.

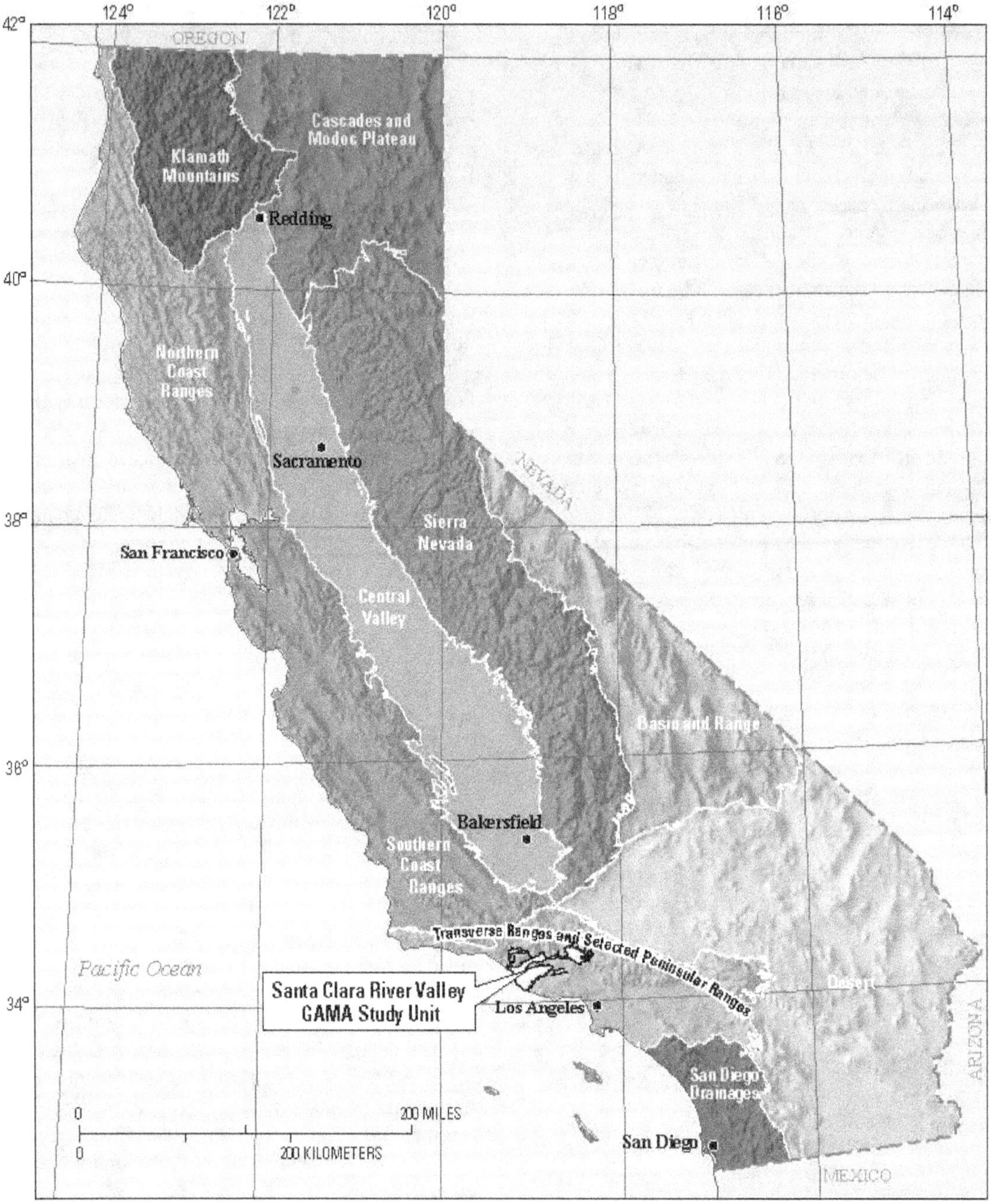

Shaded relief derived from U.S. Geological Survey National Elevation Dataset, 2006, Albers Equal Area Conic Projection, North American Datum of 1983 (NAD 83)

Provinces from Belitz and others, 2003

Figure 1. Location of the Santa Clara River Valley study unit and California hydrogeologic provinces (modified from Belitz and others, 2003), California GAMA Priority Basin Project.

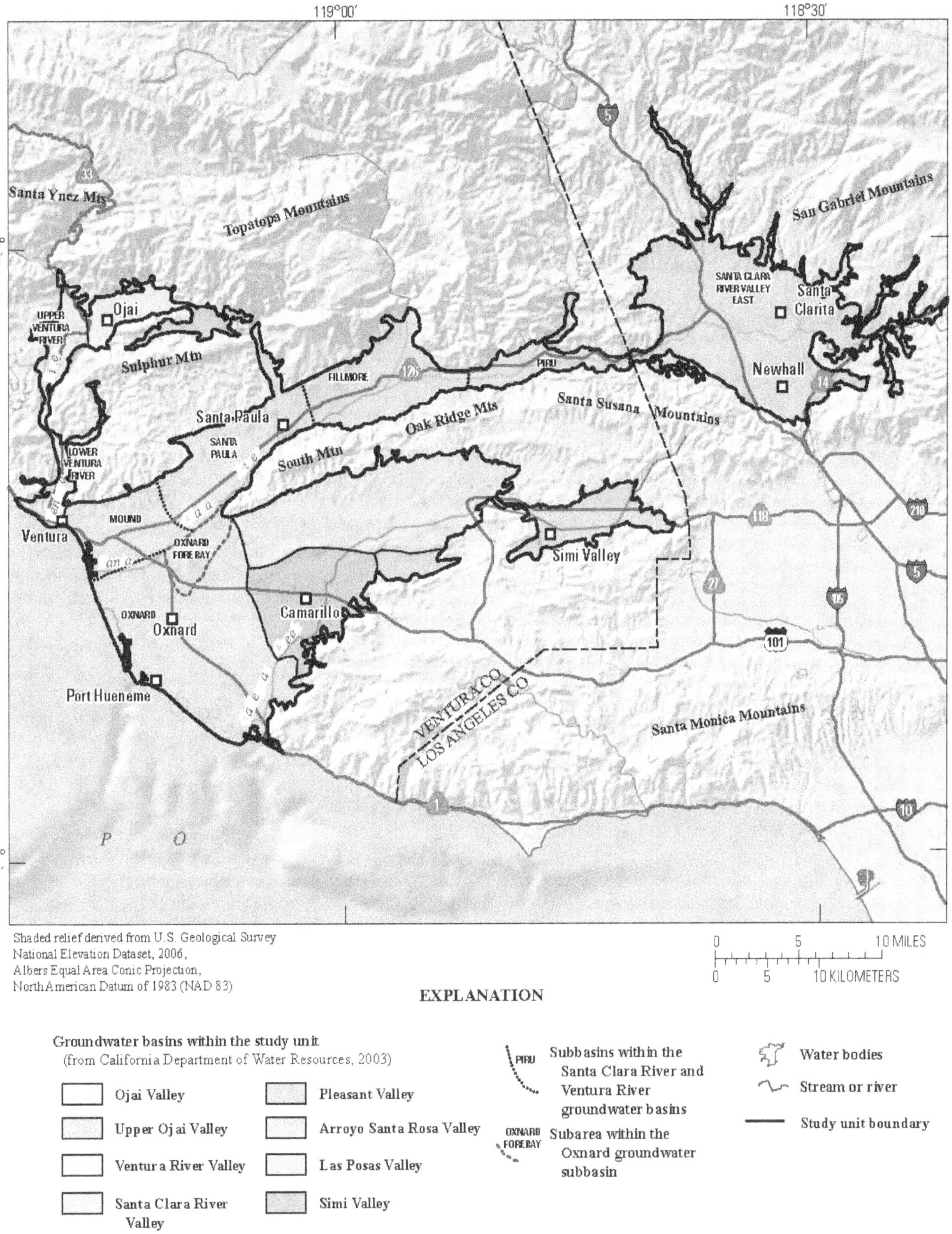

Figure 2. Study area and geographic features of the Santa Clara River Valley study unit, California GAMA Priority Basin Project.

Description of the Santa Clara River Valley Study Unit

The SCRV study unit lies in the northwestern part of the Transverse Ranges and selected Peninsular Ranges hydrogeologic province described by Belitz and others (2003) and covers about 460 mi^2 (1,192 km^2) in parts of Los Angeles and Ventura Counties, California. The SCRV study unit includes eight groundwater basins as defined by the California Department of Water Resources (2004a–2004n): Ojai Valley, Upper Ojai Valley, Ventura River Valley, Santa Clara River Valley, Pleasant Valley, Arroyo Santa Rosa Valley, Las Posas Valley, and Simi Valley. The study unit is bounded to the south by the Santa Monica Mountains and to the north by the Topatopa and Santa Ynez Mountains. The San Gabriel Mountains form the eastern boundary, and the Pacific Ocean lies to the west of the study area (fig. 2).

The land-surface altitude of the SCRV rises from sea level in the western part of the SCRV to more than 1,440 ft (440 m) in the eastern part of the Santa Clara River Valley basin. However, the land-surface altitudes of the surrounding mountains are as high as 5,800 ft (1,768 m). The climate in the SCRV is characterized as a Mediterranean climate with cool, moist winters, and dry, warm summers. The average annual precipitation ranges from 12 to 28 in/yr (30 to 71 cm/yr), with most rainfall occurring in the winter months and at high altitudes (California Department of Water Resources 2004c, 2004e, 2004i, 2004m).

Land use in the study unit is 40 percent natural, 37 percent agricultural, and 23 percent urban (fig. 3), based on the classification of USGS National Land Cover Data (Nakagaki and others, 2007). The natural land use consists of shrubs and grassland, with wetlands along the rivers and forested areas at higher altitudes. Natural land use is primarily in the Santa Clara River Valley East subbasin. Agricultural and urban land uses predominate throughout the remainder of the study unit (fig. 4). However, there is natural land use in the areas of wetlands and lagoons near the mouths of the Santa Clara River and Calleguas Creek in the western part of the study unit. The primary crops in the SCRV study unit are citrus, avocados, alfalfa, pasture, strawberries, dry beans, along with additional vegetables and subtropical fruits (California Department of Water Resources, 2001; Hanson and others, 2003). The largest urban areas in the western part of the study unit are the cities of Ventura, Oxnard, Camarillo, and Simi Valley (fig. 4). Large urban areas in the eastern part of the study unit include the cities of Newhall and Santa Clarita.

Regionally, groundwater primarily flows from the upland or elevated areas of the study unit toward the Pacific Ocean. Hanson and others (2003) developed a numerical groundwater flow model of the aquifer systems in the Santa Clara–Calleguas groundwater system, which includes the Santa Clara River Valley, Pleasant Valley, Las Posas Valley, and Arroyo Santa Rosa Valley groundwater basins, as defined by the CDWR. Hanson and others (2003) included a stratigraphic column and a related aquifer designation for geologic units used in the flow model (table 1). Public-supply and irrigation wells that withdraw water from the Santa Clara–Calleguas groundwater system are screened in the Upper- and Lower-aquifer systems. The Upper-aquifer system as defined by Turner (1975) and Hanson and others (2003) includes the Shallow, Oxnard, and Mugu aquifers. The Lower-aquifer system includes the Hueneme, Fox, and Grimes Canyon aquifers. Prior to urban and agricultural development, groundwater discharge was through evapotranspiration, discharge to rivers, and discharge to the sea through seeps in submarine canyons and cliffs (Hanson and others, 2003). After development, groundwater pumping for agricultural use accounts for the greatest amount of discharge from the aquifer system (Hanson and others, 2003), followed by municipal use. Recharge to the groundwater system is through percolation from the three main river systems—Ventura River, Santa Clara River, and Calleguas Creek—along with infiltration of direct precipitation on the landscape (Hanson and others, 2003; California Department of Water Resources, 2004a, 2004b, 2004c, 2004d, 2004e, 2004f, 2004g, 2004h, 2004i, 2004j, 2004k, 2004l, 2004m, 2004n). The source of water for streams is from runoff of precipitation, reservoir discharge, dewatering wells, and treated wastewater effluent. Percolation of irrigation-return water also contributes to groundwater recharge, but the amount of recharge from this source varies across the study unit.

The groundwater basin and subbasin boundaries in the SCRV study unit are largely controlled by faulting, folding, and deformation associated with the regional tectonic regime of the Transverse Ranges (Emery, 1960; Vedder and others, 1969; Dahlen and others, 1990; Izbicki, 1996b; Izbicki and Martin, 1997) (fig. 5). The present-day geomorphic features of the SCRV study unit are mostly a result of Middle through Late Pleistocene north-south compressional tectonics (Greene and others, 1978; Yeats and others, 1988). Uplift on the eastern part of the Oak Ridge fault and other faults have created east-west trending mountain chains (South, Oak Ridge, and Santa Susana Mountains) that separate groundwater basins to the north (Santa Clara River Valley) and south (Las Posas, Pleasant, Arroyo Santa Rosa, and Simi Valleys, fig. 5). The bedrock underlying the SCRV study unit consists of low-permeability upper Cretaceous and Tertiary sedimentary, volcanic, igneous, and metamorphic rocks. Many of the geologic formations outcropping in and around the study unit consist of marine sedimentary deposits of Tertiary age (fig. 5). The Tertiary marine rocks include the Miocene Monterey Formation, which consists of thick units of sulfur-rich organic siliceous and carbonaceous shale containing oil, tar, and metal sulfide minerals [for example, pyrite (FeS_2)] (Davis, 1961; Stankiewicz and others, 1996; Gutierrez-Alonso and Gross, 1997). The marine sedimentary formations were uplifted and became some of the source material for younger formations that make up the alluvium and aquifer material in the SCRV study unit (Hsü and others, 1980; Mangelodorf and Rullkötter, 2003).

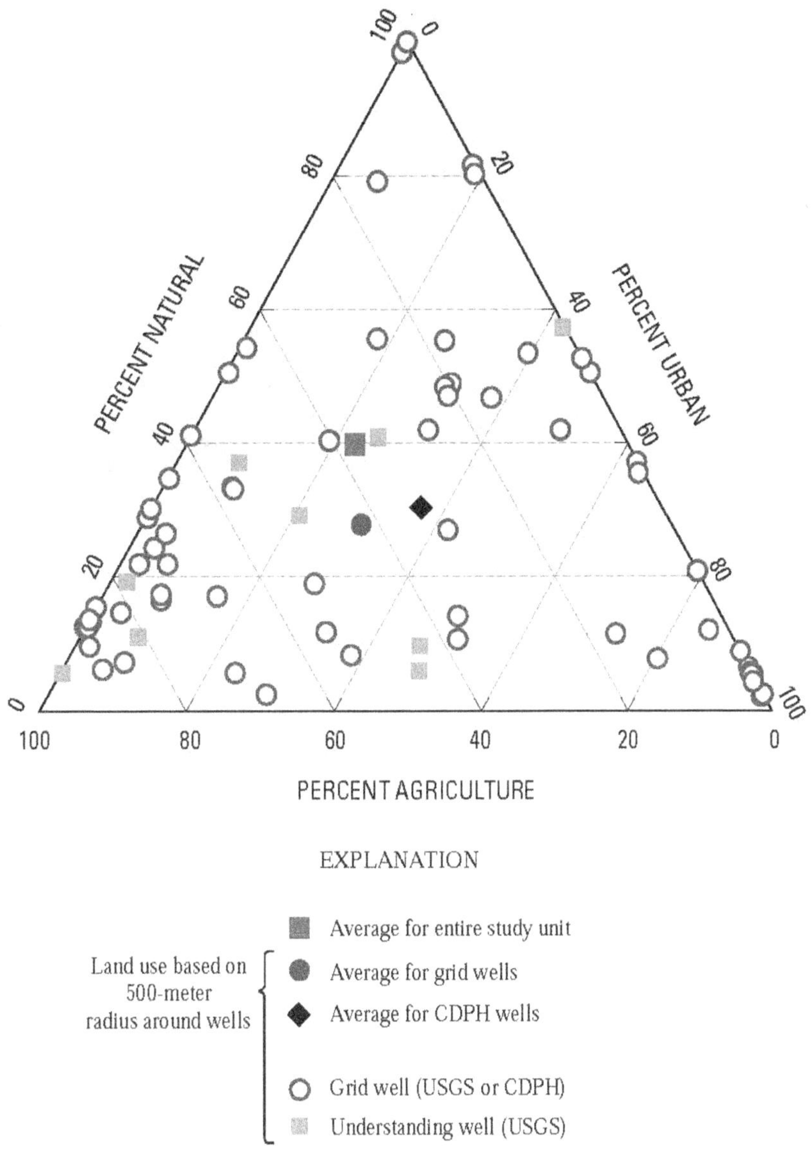

Figure 3. Proportions of urban, agricultural, and natural land use for USGS- and CDPH-grid wells, USGS-understanding wells, and other CDPH wells, Santa Clara River Valley study unit, California GAMA Priority Basin Project. (Land use was determined from USGS National Land Cover Data from Nakagaki and others, 2007.)

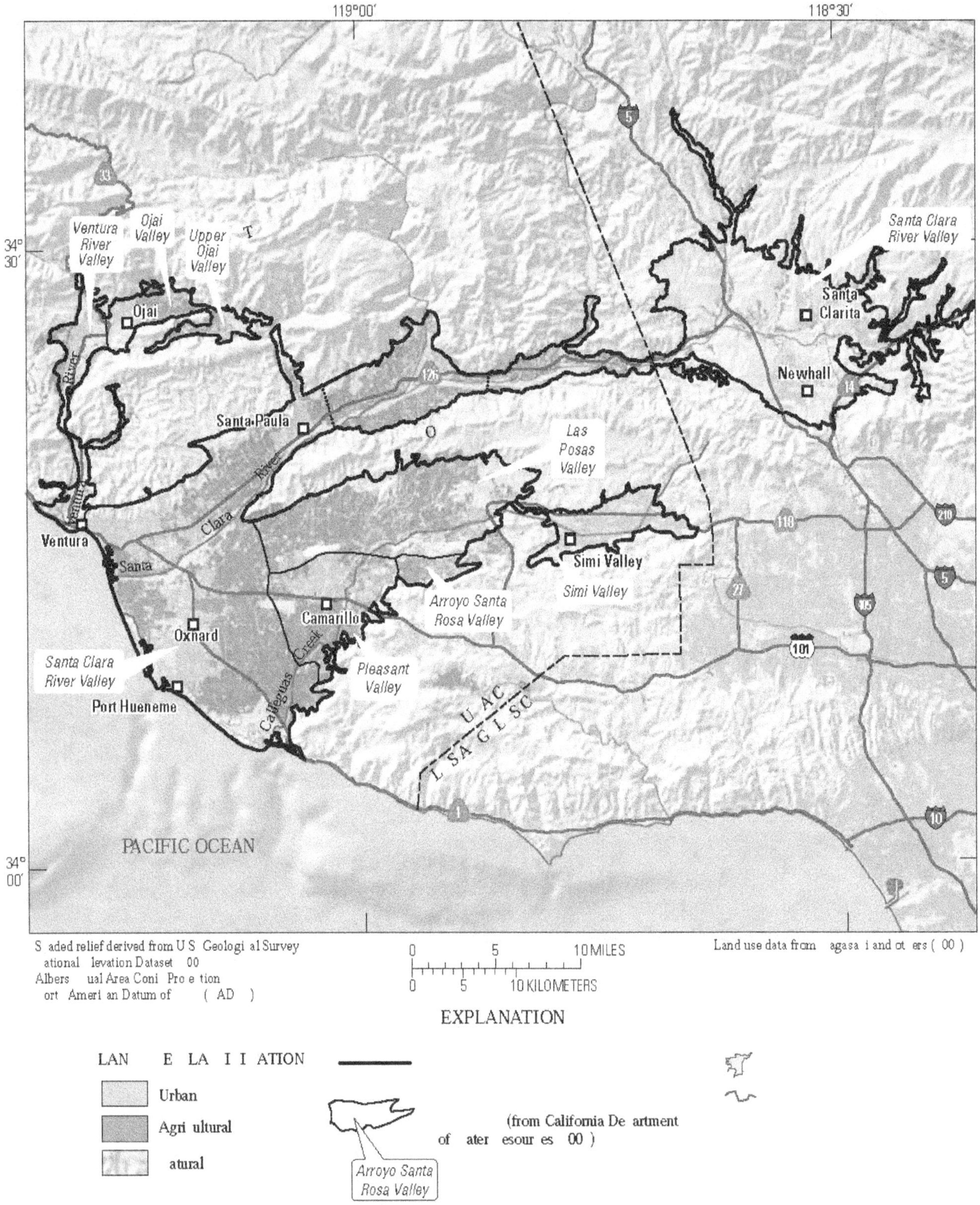

Figure 4. Land use in the Santa Clara River Valley study unit, California GAMA Priority Basin Project.

Table 1. Stratigraphic column and related designations of geologic units by source and aquifer system model layers in the groundwater and surface-water flow model of the Santa Clara–Calleguas Basin, Ventura County, California.

[Modified from Hanson and others, 2003]

Geologic era	Geologic system	Geologic series (epoch)	Weber and others (1976)	Dibblee [1]	Turner (1975) Green and others (1978) [2]	RASA [3]	Aquifer system model layers
			Lithologic units and formations		Aquifers		
Cenozoic	Quaternary	Holocene	A (Lagoonal, beach, river and flood plain deposits, artificial fill, and alluvial fan deposits)		Recent alluvial and semiperched	Shallow	Upper-aquifer system [4], layer 1
			A (Lagoonal, beach, river and flood plain deposits, and alluvial fan deposits)		Oxnard [5]		
		Late (Upper) Pleistocene [6]	O A (Lagoonal, beach, river and flood plain, alluvial fan, terrace, and marine terrace deposits)		Mugu [2]		
			[7] (Terrestrial fluvial sediments)	Saugus Formation	Hueneme	Upper Hueneme	Lower-aquifer system, layer 2
			P n [8] (Marine clays and sands and terrestrial fluvial sediments)			Lower Hueneme	
				Las Posas Sand (Marine shallow regressive sands)	Fox Canyon	Fox Canyon	
		Early (Lower) Pleistocene [6]	[8] (Marine shallow regressive sands)		Grimes Canyon [9,10]	Grimes Canyon	
			P [11] (Marine siltstones, sandstones, and conglomerates)		Formation not included in regional flow model		Formation not included in regional flow model
	Tertiary	Pliocene [6]	(Terrestrial fluvial sandstones, sandstones, and shales)				
		Miocene	T [12] (Terrestrial fluvial sandstones and fine-grained lake deposits)		Not included	Santa Margarita sandstones included in north-eastern Santa Rosa Valley	Lower-aquifer system, layer 2
			(Terrestrial and marine extrusive and intrusive, felsic-andesites to basalts)		Formation not included in regional flow model		Formation not included in regional flow model
			L T T (Marine transgressive sands and siltstones)				

Table 1. Stratigraphic column and related designations of geologic units by source and aquifer system model layers in the groundwater and surface-water flow model of the Santa Clara–Calleguas Basin, Ventura County, California.— Continued

[Modified from Hanson and others, 2003]

Geologic era	Geologic system	Geologic series (epoch)	Weber and others (1976)	Dibblee [1]	Turner (1975) Green and others (1978) [2]	RASA [3]	Aquifer system model layers
			Lithologic units and formations		Aquifers		
Cenozoic	Tertiary	Oligocene	(Terrestrial fluvial claystones and sandstones)	Formation not included in regional flow model			Formation not included in regional flow model
Cenozoic	Tertiary	Eocene	L (Marine sandstones, mudstones, and claystones)				
Cenozoic	Tertiary	Paleocene	(Terrestrial conglomerate, sandstones, and marine shales)				
Mesozoic	Upper Cretaceous		(Sandstones with shales)				

[1] Formations from Dibblee (1988, 1990a, 1990b, 1991, 1992a, 1992b, 1992c, 1992d).

[2] Perched aquifer designated in parts of the Oxnard Plain only.

[3] From Hanson and others (2003) as part of the Southern California Regional Aquifer-System Analysis Program of the U.S. Geological Survey.

[4] Shallow aquifer included in the Oxnard Plain Forebay and inland subbasins. Semiperched part of Shallow aquifer not included in remainder of Oxnard Plain.

[5] Restricted to the Oxnard Plain and Forebay by Turner (1975).

[6] Modified on the basis of ash-deposit age dates (Yerkes and others, 1987, fig. 11.2).

[7] Mapped in eastern Ventura County subbasins of Santa Paula, Fillmore, Piru, and Las Posas Valley and may be time equivalent to parts of the San Pedro and Santa Barbara Formations (Weber and others, 1976, fig. 3).

[8] Mapped in western Ventura County subbasins.

[9] San Pedro Formation everywhere in Pleasant Valley where the Santa Barbara Formation was assigned to the Grimes Aquifer.

[10] Las Posas and Pleasant Valley Basins only.

[11] Includes mud pit and claystone members.

[12] Also known as Santa Margarita Sandstone.

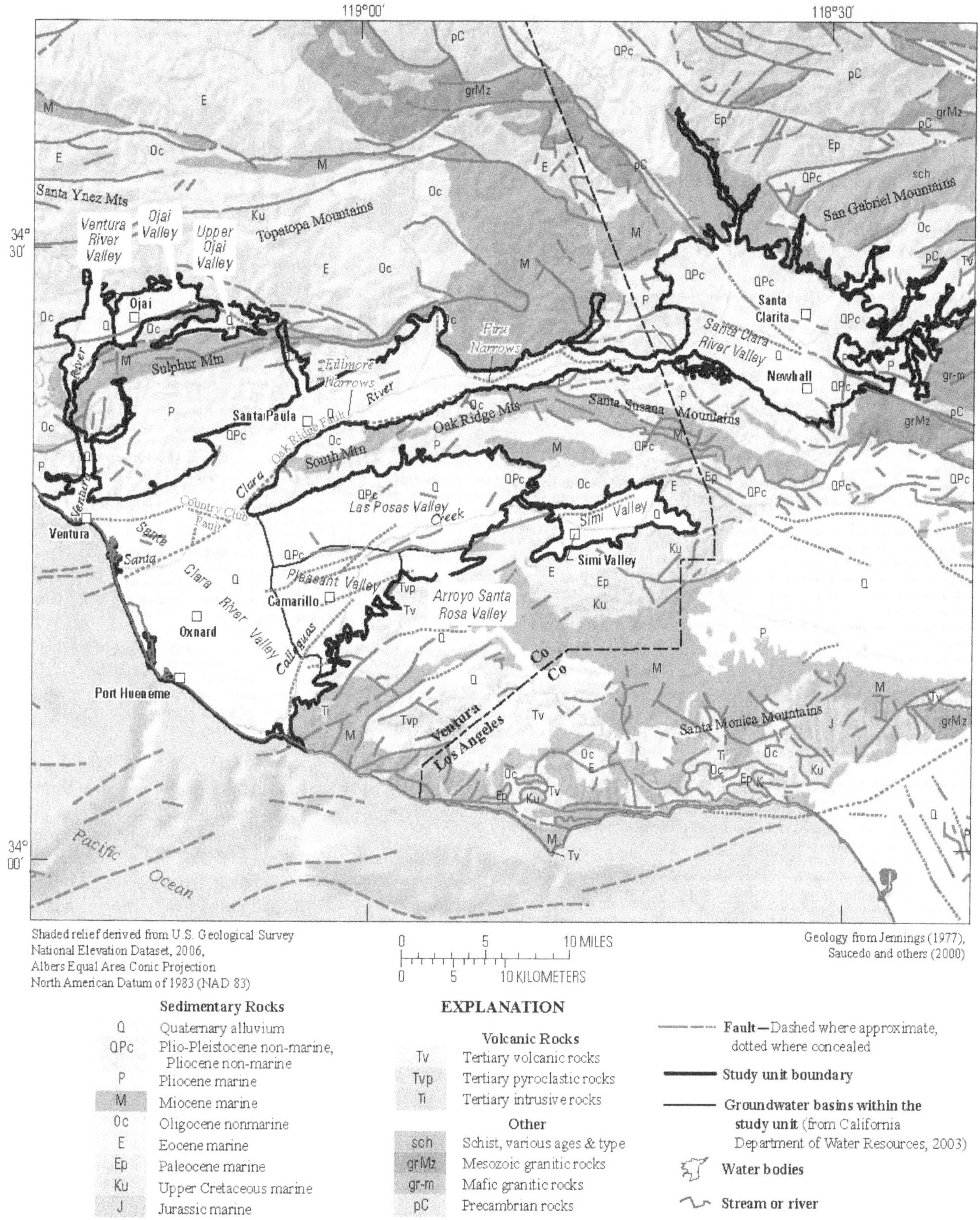
119°00′
118°30′
34°30′
34°00′
Santa Ynez Mts
Ventura River Valley
Ojai Valley
Upper Ojai Valley
Topatopa Mountains
San Gabriel Mountains
Ojai
Sulphur Mtn
Edimore Narrows
Piru Narrows
Santa Paula
Oak Ridge Fault
Oak Ridge Mts
South Mtn
Santa Clarita
Santa Clara River Valley
Newhall
Santa Susana Mountains
Country Club Fault
Las Posas Valley
Simi Valley
Ventura
Santa Clara River Valley
Pleasant Valley
Camarillo
Arroyo Santa Rosa Valley
Oxnard
Port Hueneme
Calleguas Creek
Ventura Co
Los Angeles Co
Santa Monica Mountains
Pacific Ocean
Shaded relief derived from U.S. Geological Survey National Elevation Dataset, 2006, Albers Equal Area Conic Projection North American Datum of 1983 (NAD 83)
0 5 10 MILES
0 5 10 KILOMETERS
Geology from Jennings (1977), Saucedo and others (2000)
EXPLANATION
Sedimentary Rocks
Q Quaternary alluvium
QPc Plio-Pleistocene non-marine, Pliocene non-marine
P Pliocene marine
M Miocene marine
Oc Oligocene nonmarine
E Eocene marine
Ep Paleocene marine
Ku Upper Cretaceous marine
J Jurassic marine
Volcanic Rocks
Tv Tertiary volcanic rocks
Tvp Tertiary pyroclastic rocks
Ti Tertiary intrusive rocks
Other
sch Schist, various ages & type
grMz Mesozoic granitic rocks
gr-m Mafic granitic rocks
pC Precambrian rocks
Fault—Dashed where approximate, dotted where concealed
Study unit boundary
Groundwater basins within the study unit (from California Department of Water Resources, 2003)
Water bodies
Stream or river

Figure 5. Geology of the Santa Clara River Valley study unit, California GAMA Priority Basin Project.

Ventura River Valley, Ojai Valley, and Upper Ojai Valley Basins

The primary freshwater-bearing deposits vary across the SCRV study unit. The water-bearing formations in the Upper and Lower Ventura River Valley subbasins and Ojai Valley and Upper Ojai Valley basins are alluvium of Pleistocene and Holocene age and largely are unconfined. The thickness of the alluvium typically ranges from 50 to 100 ft, although the thickness can be as much as 300 ft in parts of the Upper Ojai Valley basin. In the Lower Ventura River Valley subbasin, groundwater also is pumped from the middle Pleistocene age San Pedro Formation, which is partially confined near the mouth of the Ventura River (California Department of Water Resources, 2004d, 2004f, 2004m, 2004n).

Simi Valley and Arroyo Santa Rosa Valley Basins

The water-bearing formations in the Simi Valley and Arroyo Santa Rosa Valley groundwater basins are alluvium of Pleistocene and Holocene age and generally are unconfined (California Department of Water Resources, 2004a, 2004l). The alluvium reaches a maximum thickness of 730 ft in the Simi Valley basin but becomes shallow in the western part of the valley. The alluvium reaches a thickness of 200 ft in the Arroyo Santa Rosa Valley basin. The San Pedro Formation also is a major water-bearing unit in the Arroyo Santa Rosa Valley. The San Pedro Formation reaches a thickness of 700 ft, and is a confined aquifer in the western part of the basin.

Santa Clara River Valley Basin

The Santa Clara River Valley groundwater basin consists of six subbasins (fig. 2): Santa Clara River Valley East, Piru, Fillmore, Santa Paula, Mound, and Oxnard. The Piru, Fillmore, and Santa Paula subbasins, located along the Santa Clara River, are separated by a series of narrows created by reverse faults on the northern and southern sides of the valley. The Piru and Fillmore narrows (fig. 5) are caused by the more competent Tertiary rock reducing the width of the subbasins, causing groundwater to discharge back to the surface (Reichard and others, 1999). The Mound subbasin, also located along the Santa Clara River, is mostly separated from Santa Paula subbasin by the Country Club fault (figs. 2 and 5), which impedes groundwater flow (Reichard and others, 1999).

In the Santa Clara River Valley East subbasin, the primary water-bearing deposits are Holocene-age alluvium, Pleistocene-age terrace deposits, and the Pleistocene-age Saugus Formation (California Department of Water Resources, 2004k). The alluvial deposits are thickest (as much as 240 ft) beneath the Santa Clara River channel. Terrace deposits are located along the low-lying flanks of the foothills and the upper reaches of tributaries to the Santa Clara River but yield little groundwater. The Saugus Formation consists of two parts in the Santa Clara River Valley East subbasin. Water in the lower part is brackish and well yield is low. The upper part of the Saugus Formation contains poorly consolidated lenses of conglomerate and sandstone interbedded with sandy mudstone. The maximum depth of the upper part reaches 5,500 ft below land surface. The upper part provides much higher well yields and water that is of better quality than the lower part.

Upper- and Lower-aquifer systems (Hanson and others, 2003) can be recognized in the Piru, Fillmore, Santa Paula, and Mound subbasins. The Lower-aquifer system generally is unconfined in the Piru and Fillmore subbasins, but becomes confined in the western parts of the Santa Paula and Mound subbasins. Hanson and others (2003) identify the Hueneme aquifer of the Saugus Formation as the uppermost aquifer within the Lower-aquifer system in these four groundwater subbasins (table 1). The California Department of Water Resources (2004b, 2004e, 2004h, 2004j) identifies the primary water-bearing formation as the San Pedro Formation in the Lower-aquifer system. Yeats and others (1988) point out that the geologic formation name, "Saugus," has been used by previous authors in reference to shallow marine and non-marine deposits of Pleistocene age, typically referred to as the San Pedro Formation (Weber and others, 1973). The depths of production wells sampled as part of this study in the Piru, Fillmore, Santa Paula, and Mound subbasins range from 275 to 1,190 ft below land surface. The shallow wells are screened in the Upper-aquifer system, the deeper wells usually are screened in the Lower-aquifer system, and a few wells are screened in both aquifers.

The Upper- and Lower-aquifer systems are better defined in the Oxnard subbasin than in the Piru, Fillmore, and Santa Paula subbasins. The Upper-aquifer system primarily consists of the Pleistocene and Holocene age sands and gravels. The Lower-aquifer system consists of Pleistocene marine and non-marine deposits. The Lower-aquifer system is unconfined in the Oxnard forebay (fig. 2) located adjacent to the Santa Clara River in the northeastern part of the Oxnard subbasin and is hydraulically connected to the Upper-aquifer system. The United Water Conservation District (UWCD) recharge facilities, which are located in the Oxnard forebay, play an important role in recharging both aquifer systems of the Oxnard subbasin (United Water Conservation District, 2003). Wells sampled as part of this study are screened in the Upper-aquifer, the Lower-Aquifer or both aquifer systems.

Because of pumping along coastal areas of the Oxnard subbasin, seawater intrusion is a concern for local water management authorities (California Department Water Resources, 1965; Greene and others, 1978; Izbicki and others, 1995; Izbicki, 1996a, 1996c; United Water Conservation District, 2003; Fox Canyon Groundwater Management Agency, 2007).

Las Posas Valley and Pleasant Valley Basins

The Las Posas Valley and Pleasant Valley groundwater basins also have defined Upper- and Lower-aquifer systems. The Upper-aquifer system in these basins consists of unconfined alluvial deposits of Pleistocene and Holocene age. The thickness of these alluvial deposits reaches as much as 300 ft in the Las Posas Valley basin. A regional unconformity, which is caused by a deformation during deposition of older alluvium, exists at the base of the Upper-aquifer system separating the overlying alluvial deposits from the underlying San Pedro Formation and Santa Barbara Formation that make-up the Lower-aquifer system (Turner, 1975; Greene and others, 1978). This unconformity is more distinct in the eastern part of the Las Posas Valley basin. Alluvial deposits in Pleasant Valley are not a major source of groundwater (California Department of Water Resources, 2004i). The Lower-aquifer system is composed of the Hueneme, Fox Canyon, and Grimes Canyon aquifers (table 1). The Hueneme Aquifer consists of the upper part of the San Pedro and Saugus Formations; the Fox Canyon aquifer consists of the lower part of the San Pedro Formation, which is composed of coarse gravel 100–200 ft thick in Las Posas Valley basin and as much as 300 ft thick in Pleasant Valley basin (Hanson and others, 2003; California Department of Water Resources, 2004c, 2004i). The Santa Barbara Formation, deposited during the early Pleistocene, underlies the San Pedro Formation; and is composed of shallow marine sand. The Grimes Canyon aquifer is composed of the upper part of the Santa Barbara Formation, which is as much as 1,000 ft thick in the Las Posas Valley basin (Hanson and others, 2003; California Department of Water Resources, 2004c). Most of the wells sampled in this part of the study unit are screened from about 500 to 1,000 ft deep.

Methods

Methods used for the GAMA Priority Basin Project were selected to achieve the following objectives: (1) design a sampling plan suitable for statistical analysis, (2) combine CDPH data with data collected in 2007 by the USGS for assessing water quality, (3) determine proportions of the primary aquifers that have high, moderate, and low concentrations for constituent classes and individual constituents of interest, (4) identify constituents of interest to be discussed further, (5) compile and classify relevant ancillary data to identify relations of potential explanatory factors to water quality, and, (6) investigate statistical relations between potential explanatory factors and water quality.

The *status assessment* was designed to provide a spatially unbiased assessment of untreated groundwater quality in the primary aquifer systems. The primary metric for defining groundwater quality in this report is *relative-concentration*, which compares groundwater chemical concentrations to regulatory and non-regulatory benchmarks used to evaluate drinking-water quality. Constituents are included or not included in the assessment on the basis of relative-concentration criteria. Groundwater-quality data collected by the USGS for the GAMA Program and data compiled in the CDPH database are used in the *status assessment*. Two statistical approaches based on spatially unbiased grids with equal-area cells in each study area are used to calculate aquifer-scale proportions.

The *understanding assessment* was designed to evaluate the natural and human factors that affect groundwater quality at the study-unit level. A finite set of potential explanatory factors were analyzed in relation to constituents of interest to place the observed water quality within the context of physical and chemical processes. Statistical tests were used to identify significant correlations between the constituents of interest and potential explanatory factors.

Status Assessment Methods

The *status assessment* included two primary steps. First, water-quality data were normalized to their respective water-quality benchmarks by calculating their relative-concentrations (Toccalino and others, 2004; Toccalino and Norman, 2006). Second, aquifer-scale proportions were determined for categories of "high," "moderate," and "low" relative-concentrations by using two approaches: (1) grid-based and (2) spatially weighted (Belitz and others, 2010). The "grid-based" approach uses one well per cell to represent groundwater quality—water-quality data are from wells sampled by the USGS, supplemented with data from selected wells in the CDPH database. The spatially weighted approach uses data for wells sampled by the USGS and all wells in the CDPH database, and weights each well such that each grid cell contributes equally to represent groundwater quality. Results for the two approaches were compared, and results from the preferred approach were used to identify constituents of interest for further discussion.

Relative-Concentrations and Water-Quality Benchmarks

Concentrations of constituents are presented as relative-concentrations in the *status assessment*:

$$\text{Relative-concentration} = \frac{\text{Sample concentration}}{\text{Water-quality benchmark concentration}}.$$

Relative-concentrations provide a means to relate concentrations of constituents in groundwater samples to water-quality benchmarks. Relative-concentrations less than 1.0 indicate sample concentrations less than the benchmark; relative-concentrations greater than 1.0 indicate sample concentrations greater than the benchmark. The use of relative-concentrations also normalizes a wide range of concentrations for different constituents to a common scale relative to benchmarks. Toccalino and others (2004), Toccalino and Norman (2006), and Rowe and others (2007) used the ratio of measured concentration to a benchmark [either MCLs or Health-Based Screening Levels (HBSLs)] and defined this ratio as the Benchmark Quotient. Relative-concentrations used in this report are equivalent to the Benchmark Quotient reported by Toccalino and others (2004) for constituents with water-quality benchmarks. HBSLs were not used in this report because HBSLs are not currently used as benchmarks by California drinking-water regulatory agencies. Relative-concentrations were computed only for compounds with water-quality benchmarks; therefore, constituents lacking water-quality benchmarks were not included in the status assessment.

Regulatory and non-regulatory water-quality benchmarks apply to water that is served to the consumer, not to untreated groundwater. However, to provide context for the water-quality results, concentrations of constituents measured in the untreated groundwater were compared with regulatory and non-regulatory human-health-based water-quality benchmarks established by the U.S. Environmental Protection Agency (USEPA) and CDPH (U.S. Environmental Protection Agency, 2006; California Department of Health Services, 2007). The human-health benchmarks used include MCLs, notification levels (NLs), health advisory levels (HALs), action levels (ALs), and risk-specific dose (1 in 100,000 lifetime risk of cancer, RSD5-US). Non-regulatory benchmarks set for aesthetic concerns, secondary maximum contaminant levels defined by CDPH and USEPA (SMCL-CA and SMCL-US), also were used. For a constituent with multiple types of benchmarks, the benchmark used for calculation of relative-concentration was selected according to the following order of priority: regulatory human-health (MCL and AL), non-regulatory aesthetic (SMCL), and non-regulatory human-health (in the order NL-CA, HAL-US, and RSD5-US). For the regulatory human-health benchmarks, Federal benchmark levels were used unless the California levels were lower. California public health goals were not used in this report. Additional information on the types of benchmarks and the benchmarks for all constituents analyzed is provided by Montrella and Belitz (2009).

Relative-concentrations were classified into high, moderate, and low categories.

Category	Relative-concentrations for organic constituents	Relative-concentrations for inorganic constituents
High	> 1	> 1
Moderate	> 0.1 and ≤ 1	> 0.5 and ≤ 1
Low	≤ 0.1	≤ 0.5

A relative-concentration of 0.1 was used as a threshold between low and moderate values of organic and special-interest constituents compared with a relative-concentration of 0.5 for inorganic constituents. Organic and special-interest constituents, which generally are anthropogenic and uncommon in groundwater, are usually less prevalent and have smaller relative-concentrations than inorganic constituents. The USEPA also established a relative-concentration of 0.1 of the regulatory benchmark as a threshold concentration so that the agency would be notified if the presence of a pesticide in surface water or groundwater equals or exceeds that threshold (U.S. Environmental Protection Agency, 1998). In contrast, inorganic constituents typically occur naturally at concentrations that could be greater than 0.1 of regulatory benchmarks; consequently, it would be difficult to identify the highest-priority inorganic constituents that may have elevated concentrations greater than background levels if a relative-concentration of 0.1 were used as the threshold between moderate and low relative-concentrations. Therefore, a relative-concentration of 0.5 was used as a threshold between low and moderate values of inorganic constituents, rather than 0.1 as was used for the organic and special-interest constituents.

Design of Sampling Network for Status Assessment

The wells selected for sampling by the USGS in this study were selected to provide a statistically unbiased, spatially distributed set of wells for the assessment of the quality of groundwater in the primary aquifers. Water-quality data from the USGS-grid wells were supplemented with data from selected wells from the CDPH database [CDPH-grid wells—discussed in more detail in the California Department of Public Health (CDPH)-Grid Well Selection section] to obtain more complete grid coverage and to include constituents that were not analyzed for in every USGS-grid well. These data were used to assess the proportions of the primary aquifer system that have high, moderate, and low relative-concentrations.

The primary data used for the grid-based calculations of aquifer-scale proportions were data from wells sampled by the GAMA Priority Basin Project. Detailed descriptions of the methods used to identify wells for sampling are given in Montrella and Belitz (2009). Briefly, the study unit was divided into 48 equal-area grid cells of about 10 mi^2 (25 km^2) each (fig. 6); one well was randomly selected to represent each cell (Scott, 1990). Wells were selected from the population of wells in statewide databases maintained by CDPH and USGS. If a grid cell did not contain accessible CDPH wells, then commercial, irrigation, or domestic wells with perforation intervals at similar depths as the CDPH wells were considered for sampling. One USGS-grid well was sampled in 42 of the 48 grid cells (fig. 6). The six grid cells where samples were not collected had few, if any, wells, and (or) permission to sample was not granted for wells in those cells. The 42 USGS-grid wells were sampled during the period April through June 2007 and included 27 CDPH wells, 12 irrigation wells, 2 dewatering wells, and 1 domestic well. USGS-grid wells in the study unit were numbered in the order of sample collection and with the prefix SCRV (fig. A1*A*). The USGS-grid wells were sampled by the USGS for the GAMA Priority Basin Project but are owned by other organizations or individuals.

Samples collected from USGS-grid wells were analyzed for 232 to 374 constituents (table 2). VOCs, pesticides, potential wastewater indicator compounds, perchlorate, noble gases, tritium, and stable isotopes of hydrogen and oxygen were analyzed in water samples from all wells. Nutrients, dissolved organic carbon, isotopes of nitrogen and oxygen in nitrate, major and minor ions, trace elements, and redox species were analyzed in samples from many wells. Gasoline oxygenates, pharmaceuticals, additional pesticides, carbon isotopes, isotopes of chlorine and bromine, and radiochemical constituents were analyzed in samples from some wells. The collection, analysis, and quality-control data for the analytes listed in table 2 are described by Montrella and Belitz (2009).

Table 2. Analytes and numbers of wells sampled for each analytical schedule, Santa Clara River Valley study unit, GAMA Priority Basin Project, April–June 2007.

	Schedule		
	Fast	**Intermediate**	**Slow**
Total number of wells	27	17	9
Number of grid wells sampled	26	11	5
Number of understanding wells sampled	1	6	4
Analyte groups	**Number of constituents**		
Dissolved oxygen, pH, specific conductance, temperature	4	4	4
Volatile organic compounds (VOCs) [1]	85	85	85
Pesticides and degradates	63	63	63
Potential wastewater-indicator compounds	69	69	69
Perchlorate	1	1	1
Noble gases and tritium [2]	7	7	7
Tritium [3]	1	1	1
Hydrogen and oxygen isotopes of water [4]	2	2	2
Nutrients		5	5
Dissolved organic carbon		1	1
Nitrogen and oxygen isotopes of nitrate		2	2
Major and minor ions, and trace elements		35	35
Arsenic, chromium, and iron species		6	6
Alkalinity, turbidity			2
Gasoline oxygenates [5]			3
Pharmaceuticals [4]			14
Polar pesticides and degradates [6]			59
Carbon isotopes			2
Chlorine and bromine isotopes [4]			2
Radon-222			1
Radium isotopes			2
Gross alpha and beta radioactivity			4
Microbial constituents [4]			4
Sum:	232	281	374

[1] Includes 10 constituents classified as fumigants or fumigant synthesis byproducts.

[2] Analyzed at Lawrence Livermore National Laboratory, Livermore, California.

[3] Analyzed at U.S. Geological Survey Stable Isotope and Tritium Laboratory, Menlo Park, California.

[4] Not discussed in this report.

[5] Does not include five constituents in common with VOCs.

[6] Does not include four constituents in common with pesticides and degradates.

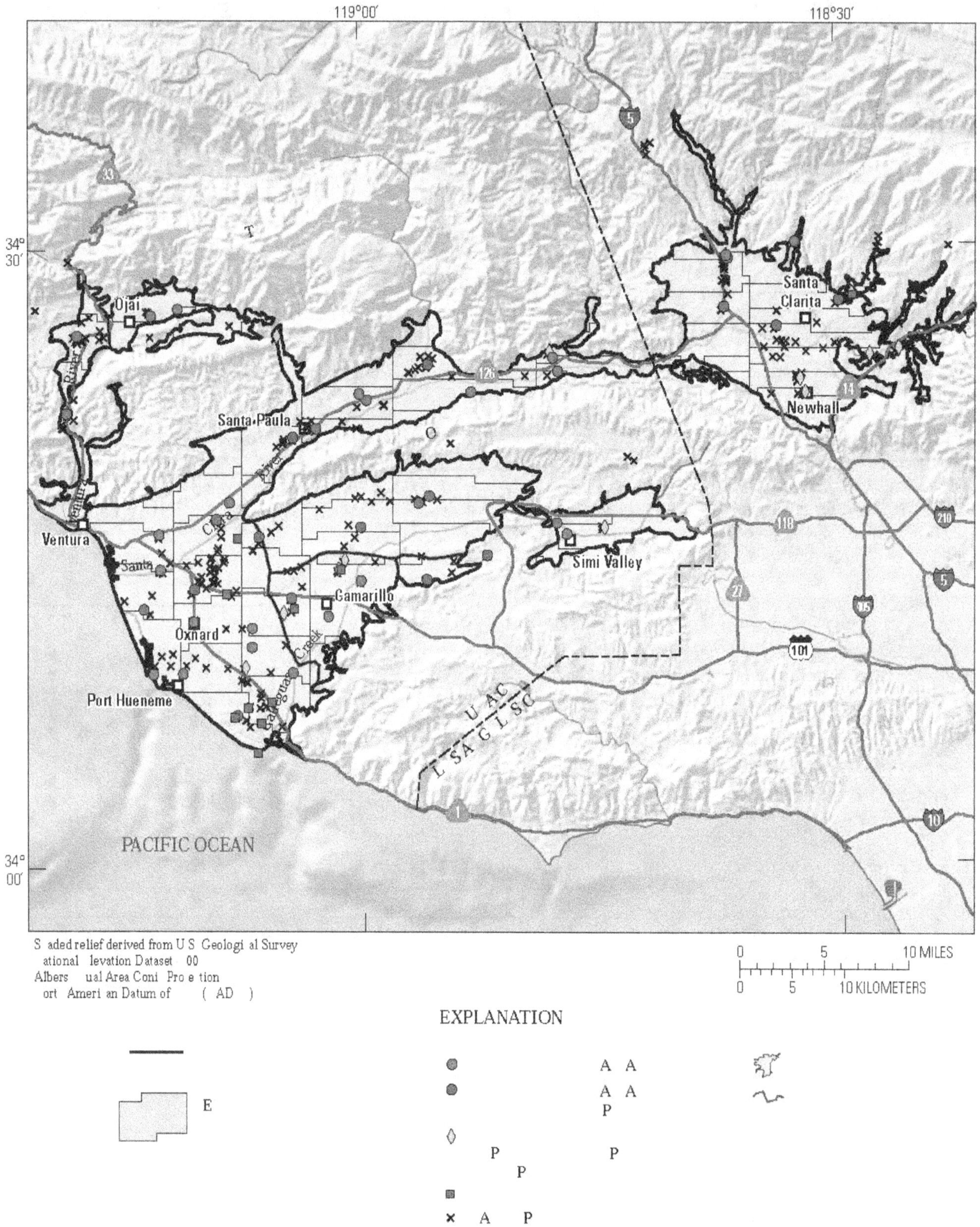

Figure 6. Location of study-unit grid cells, USGS-grid and USGS-understanding wells, California Department of Public Health (CDPH) wells used to supplement inorganic constituents (CDPH-grid wells), and other CDPH wells located in the Santa Clara River Valley study unit, California GAMA Priority Basin Project.

California Department of Public Health (CDPH)-Grid Well Selection

Data for VOCs, pesticides, and perchlorate were collected at all 42 USGS-grid wells. The USGS-grid-well data included more VOC and pesticide constituents, and reporting levels were lower, than reporting levels from the CDPH database. Therefore, CDPH data for these constituents were not used to supplement USGS-grid well data for the *status assessment*.

Samples to be analyzed for inorganic constituents were collected from 16 of 42 USGS-grid wells. Because the GAMA Priority Basin Project did not collect a complete suite of inorganic constituents for all grid cells, the CDPH database was used to provide data for inorganic constituents for the cells that lacked this data (table 3). In addition, the GAMA Priority Basin Project was not able to sample wells in six of the grid cells. CDPH wells were selected to represent as many of these grid cells as possible. CDPH wells that were selected to supplement USGS-grid wells are referred to as "CDPH-grid" wells. The approach used to identify suitable CDPH wells is described in appendix A. Briefly, the first choice was to use CDPH data from the same well as the USGS-grid well ("DG" CDPH-grid wells; fig. A1*B*; table A1). If the DG well did not have all needed data, then a second well in the same grid cell was randomly selected from the subset of CDPH wells with data ("DPH" CDPH-grid wells; fig. A1*B*; table A1). No more than one "DPH" CDPH-grid well was selected for each cell. Combining data from CDPH-grid wells with data from USGS-grid wells produced inorganic data for 47 of the 48 cells. All other CDPH wells with data from the current period (November 1, 2003, through October 31, 2006) not selected to be CDPH-grid wells are referred to as "CDPH-other" wells. Comparisons of data from USGS and CDPH wells to assess the validity of using these different sources in combination are presented in appendix B.

Identification of Constituents of Interest

The GAMA Priority Basin Project used monitoring data in the CDPH database along with newly collected data for characterization of the groundwater resource. The statewide CDPH database contains data for regulated constituents that have water-quality benchmarks. Although other organizations also collect water-quality data, the CDPH database is the only database containing statewide public-supply well water-quality data. Data for some constituents, including VOCs, pesticides, inorganic constituents, and radioactive constituents, are available from both the USGS-GAMA and CDPH databases. However, more VOCs and pesticides were analyzed for by the USGS Priority Basin Project than were available from the CDPH database (table 4). In addition, laboratory reporting levels (LRLs) for USGS-GAMA analyses typically were one to two orders of magnitude lower than the method

Table 3. Inorganic constituents and number of grid wells per constituent, Santa Clara River Valley study unit, GAMA Priority Basin Project.

[GAMA, Groundwater Ambient Monitoring and Assessment Program; CDPH, California Department of Public Health; MCL-US, U.S. Environmental Protection Agency (USEPA) maximum contaminant level; MCL-CA, CDPH maximum contaminant level; SMCL-CA, CDPH secondary maximum contaminant level; SMCL-US, USEPA secondary maximum contaminant level; NL-CA, CDPH notification level; AL-US, USEPA action level; HAL-US, USEPA lifetime health advisory; USGS, U.S. Geological Survey]

Constituent	Benchmark type	Number of grid wells sampled by GAMA	Number of grid wells selected from CDPH
Nutrients with health-based benchmarks			
Ammonia as nitrogen	HAL-US	16	4
Nitrite as nitrogen	MCL-US	16	26
Nitrate as nitrogen	MCL-US	16	30
Trace elements and minor ions with health-based benchmarks			
Aluminum	MCL-CA	16	29
Antimony	MCL-US	16	28
Arsenic	MCL-US	16	28
Barium	MCL-CA	16	27
Beryllium	MCL-US	16	28
Boron	NL-CA	16	17
Cadmium	MCL-US	16	28
Chromium	MCL-CA	16	29
Copper	AL-US	16	29
Lead	AL-US	16	20
Molybdenum	HAL-US	16	0
Nickel	MCL-CA	16	28
Selenium	MCL-US	16	28
Strontium	HAL-US	16	0
Thallium	MCL-US	16	28
Vanadium	MCL-US	16	13
Fluoride	MCL-CA	16	29
Trace elements and major ions with secondary maximum contaminant levels			
Iron	SMCL-CA	16	29
Manganese	SMCL-CA	16	29
Silver	SMCL-CA	16	28
Zinc	SMCL-US	16	29
Chloride	SMCL-CA	16	29
Sulfate	SMCL-CA	16	29
Total dissolved solids	SMCL-US	16	30
Radioactive constituents with health-based benchmarks			
Gross alpha radioactivity	MCL-US	5	24
Gross beta radioactivity	MCL-US	5	9
Radon-222	MCL-US	5	1
Radium-226 plus -228	MCL-US	5	21
Uranium	MCL-US	16	19

detection levels (MDLs) used for analyses compiled by CDPH (table 4). Thus, the newly acquired GAMA data were selected to enhance the CDPH data by providing a larger number of analytes and lower laboratory reporting levels than are found in the CDPH database. Both datasets are used in the status assessment and *understanding assessment*.

The CDPH database contains more than 235,000 historical records from about 340 wells in the SCRV, necessitating targeted retrievals to effectively access the water-quality data. CDPH data were used with USGS-grid data to identify constituents in the SCRV study unit that were detected at concentrations greater than water-quality benchmarks in any sample during the period of record. Data were retrieved from the CDPH database for samples from all wells located in the SCRV study unit for the full period of record (March 5, 1984, to October 31, 2006). Constituent concentrations were identified as historically high (table 5) if (1) concentrations were high (greater than benchmarks) at any time during the full period of record (March 5, 1984–October 31, 2006) and (2) concentrations were not high in the most recent 3-year period of CDPH data (November 1, 2003, through October 31, 2006, hereinafter referred to as current period) or in the USGS-grid data. These constituents do not reflect current conditions on which the status assessment is based.

Of the more than 300 constituents analyzed in the SCRV study unit, only those of greatest importance to water quality in the primary aquifer systems are discussed in this report. Constituents examined in the *status assessment* include:

1. Constituents that had high or moderate relative-concentrations in the CDPH database for samples collected during the 3-year period (November 1, 2003–October 31, 2006) prior to USGS-GAMA well sampling,
2. Constituents that had high or moderate relative-concentrations in the USGS-grid wells or USGS-understanding wells sampled April-June 2007 (discussed later in this report) or,
3. Organic constituents that were detected in greater than 10 percent of the USGS-grid wells, even if all relative-concentrations were low, because of their prevalence in the aquifer.

The *status assessment* particularly focused on any constituent with aquifer-scale proportions that had high values in greater than 2 percent of the primary aquifer system.

Estimation of Aquifer-Scale Proportions

Aquifer-scale proportions are defined as the percentage of the area (rather than the volume) of the primary aquifer system with concentrations greater than or less than specified thresholds relative to regulatory or aesthetic benchmarks. Two statistical approaches were selected to evaluate the proportions of the primary aquifer system (Belitz and others, 2010) in the SCRV study unit with high, moderate, or low relative-concentrations of constituents relative to benchmarks:

- Grid-based: One value per grid cell from either USGS-grid or CDPH-grid wells was used to represent the primary aquifer system. The proportion of the primary aquifer system with high relative-concentrations was calculated by dividing the number of wells (cells) represented by a high relative-concentration for a particular constituent by the total number of wells (grid cells) with data for that constituent (see appendix C for details of methods). Proportions of moderate and low relative-concentrations were calculated similarly. Confidence intervals for grid-based detection frequencies of high concentrations were computed using the Jeffreys interval for the binomial distribution (Brown and others, 2001). Although the grid-based estimate is spatially unbiased, the grid-based approach may not identify constituents that are present at high relative-concentrations in small proportions of the primary aquifers.
- Spatially weighted: All available data from the following sources were used to calculate the aquifer-scale proportions—all CDPH wells in the study unit (most recent analysis from each well with data for that constituent during the current period, November 1, 2003, to October 31, 2006), USGS-grid wells, and USGS-understanding wells with perforation depth intervals representative of the primary aquifer system. USGS-understanding wells that were monitoring wells were excluded because these wells were perforated at shallower or deeper depths than wells typically used for public supply in the areas in which the monitoring wells were located. For the spatially weighted approach, proportions were computed on a cell-by-cell basis (Isaaks and Srivastava, 1989) rather than as an average of all wells. The proportion of high relative-concentrations for each constituent for the primary aquifers was computed by (1) computing the proportion of wells with high relative-concentrations in each grid-cell; and (2) averaging together the grid-cell proportions computed in step (1) (see appendix C for details of methods). Similar procedures were used to calculate the proportions of moderate and low relative-concentrations of constituents. The resulting proportions are spatially unbiased (Isaaks and Srivastava, 1989).

The raw detection frequencies of wells with high relative-concentrations for constituents calculated by using the same data that were used for the spatially weighted approach are provided for reference in this report, but were not used to assess aquifer-scale proportions. These raw detection frequencies are not spatially unbiased because the wells in the CDPH database are not uniformly distributed. Consequently, high relative-concentrations in spatially clustered wells in a particular area representing a small part of the primary aquifers could be given a disproportionately high weight compared to spatially unbiased methods.

Table 4. Comparison of number of compounds and median method-detection levels or laboratory reporting levels by type of constituent for data stored in the California Department of Public Health (CDPH) database and data collected by the Santa Clara River Valley study unit, GAMA Priority Basin Project, April through June 2007.

[GAMA, Groundwater Ambient Monitoring and Assessment Program; MDL, method-detection level; LRL, laboratory reporting level; mg/L, milligrams per liter; µg/L, micrograms per liter; pCi/L, picocuries per liter; ssL_c, sample-specific critical level; nc, not collected]

Constituent type	CDPH		GAMA		Concentration or activity units
	Number of compounds	Median MDL	Number of compounds	Median LRL	
Volatile organic compounds (including fumigants) plus gasoline oxygenates	83	0.5	88	0.08	µg/L
Pesticides plus degradates	74	1.0	122	0.009	µg/L
Pharmaceuticals	nc	nc	14	[2]0.027	µg/L
Perchlorate	1	4.0	1	0.5	µg/L
Trace elements	20	8.0	24	0.16	µg/L
Radioactive constituents (ssL_c)	5	1.0	8	[3]1.0	pCi/L
Nutrients, major and minor ions	14	Unknown[1]	17	0.06	mg/L

[1] Median MDL for nutrients is 0.4 mg/L; MDLs for major and minor ions were not available.

[2] Value reported is a median MDL.

[3] Value reported is a median ssL_c for eight radioactive constituents collected and analyzed by GAMA.

Table 5. Constituents in wells in the California Department of Health (CDPH) database at historically high concentrations from March 5, 1984, to October 30, 2003, Santa Clara River Valley study unit, California, GAMA Priority Basin Project.

[A high analysis is defined as a concentration that is greater than human-heath benchmark for that constituent. GAMA, Groundwater Ambient Monitoring and Assessment Program; MCL-US; U.S. Environmental Protection Agency maximum contaminant level; MCL-CA, CDPH maximum contaminant level; mg/L, milligrams per liter; µg/L, micrograms per liter]

Constituent	Number of wells with historical data	Benchmark type	Benchmark value	Unit	Number of wells with at least one historically high value	Date of most recent high value
Trace elements						
Antimony	237	MCL-US	6	µg/L	1	10/01/1997
Cadmium	245	MCL-US	5	µg/L	2	07/12/1989
Chromium	245	MCL-CA	50	µg/L	2	12/12/1989
Mercury	244	MCL-US	2	µg/L	1	06/07/1999
Major and minor ions						
Fluoride	248	MCL-CA	2	mg/L	8	02/05/2001
Fumigants						
Dibromochloropropane (DBCP)	227	MCL-US	0.2	µg/L	2	01/14/1997
Dibromoethane (EDB)	226	MCL-US	0.05	µg/L	1	09/04/1992
Other organics						
Carbon tetrachloride	247	MCL-CA	0.5	µg/L	1	10/04/1991
Bis(2-ethylhexyl)phthalate	111	MCL-US	5	µg/L	4	10/01/1996

Aquifer-scale proportions discussed in this report primarily were estimated using the grid-based approach, and secondarily using the spatially weighted approach. The grid-based aquifer-scale proportions were used unless the spatially weighted proportions were significantly different. Significantly different results were defined as follows:

1. If the aquifer proportion for the high category was zero by using the grid-based approach and non-zero by using the spatially weighted approach, then the result from the spatially weighted approach was used. This situation can arise when the concentration of a constituent is high in a small fraction of the aquifer.
2. If the grid-based aquifer proportion for the high category was non-zero, then the 90 percent confidence interval (based on the Jeffreys interval for the binomial distribution, Brown and others, 2001) was used to evaluate the difference. If the spatially weighted proportion was within the 90 percent confidence interval, then the grid-based proportion was used. If the spatially weighted proportion was outside the 90 percent confidence interval, then the spatially weighted proportion was used.

Aquifer-scale proportions for the moderate and low categories primarily were determined from the grid-based estimates because for some constituents the reporting levels for analyses in CDPH were too high to distinguish between moderate and low relative-concentrations.

Aquifer-scale proportions of high relative-concentrations also were determined for classes of constituents. The classes of organic constituents for which aquifer-scale proportions were calculated include trihalomethanes, solvents, and pesticides. The classes of inorganic constituents with human-health benchmarks for which aquifer-scale proportions were calculated include trace elements, radioactive constituents, and nutrients. Classes of inorganic constituents with aesthetic benchmarks, for which aquifer-scale proportions were calculated, include major and minor ions (which include sulfate and chloride), total dissolved solids (TDS), and trace elements (manganese and iron).

Understanding Assessment Methods

The purpose of the understanding assessment is to place the observed groundwater-quality data into a physical and chemical context. A finite set of potential explanatory factors, including land use, well depth, depth to top-of-perforations, groundwater age classes, and geochemical-condition indicators, were analyzed in relation to constituents of interest. Statistical tests were used to identify significant correlations between the constituents of interest and potential explanatory factors. Selected correlated data that were most valuable for improving understanding of factors affecting water quality are shown graphically.

The USGS-GAMA data include hydrologic tracers and geochemical indicators that are not regulated water-quality constituents with human-health benchmarks. These constituents are important for understanding groundwater quality and are discussed for that purpose in this report.

USGS-Understanding Wells

For the *understanding assessment*, the grid-based data from USGS- and CDPH-grid wells were further supplemented with information from USGS-understanding wells for the purpose of assessing relations of selected water-quality constituents to explanatory factors (fig. A1*A*). Eleven wells (USGS-understanding wells) were selected for sampling by the USGS to increase the data density in several areas to assess spatial changes in water quality. The USGS-understanding wells sampled in the SCRV study unit were numbered in the order of collection, with a prefix modified from those used for the USGS-grid wells (SCRVU). These wells were selected on the basis of two design objectives.

First, USGS-understanding wells were sampled to assess changes in water quality with depth in the primary aquifer system. Four depth-dependent groundwater samples were collected from one long-screened understanding well, (SCRVU-04) located in the Pleasant Valley basin, by using sampling methods described by Izbicki (2004). A sample of the surface discharge of well SCRVU-04 was collected approximately 2 months before the depth-dependent sampling. A velocity log was collected prior to sample collection to determine which parts of the perforated interval in SCRVU-04 were contributing most of the flow into the well. Discussion of the results of the depth-dependent sampling is beyond the scope of this report.

Second, USGS-understanding wells were selected to identify changes in water quality in near-coastal aquifers in the Oxnard subbasin and Pleasant Valley basin. Previous investigations have identified elevated concentrations of total dissolved solids in groundwater in parts of these basins, partially caused by both seawater intrusion and the underlying aquifer material (California Department of Water Resources, 1965; Izbicki, 1996b). In an attempt to differentiate among the potential sources of salinity, selected USGS-grid and USGS-understanding wells were sampled for chlorine and bromine isotopes, which are constituents that may aid in identifying sources of salinity (table 2). USGS-understanding wells were selected to characterize source waters that may mix in the aquifers of the Oxnard subbasin. These source waters are referred to as “end members” because water samples representing mixtures tend to plot along mixing lines bounded by the “end member” compositions. Targeted “end members” include (1) an understanding well located adjacent to the UWCD Saticoy recharge facility, representing modern recharge (SCRVU-03), (2) deep wells representing the Lower-aquifer system (SCRVU-01, -02, -04), and (3) a shallow monitoring well adjacent to the coast, representing seawater composition (SCRVU-10). Discussions of the salinity source characterization using isotopic and other geochemical tracers are beyond the scope of this report.

Statistical Analysis

Non-parametric statistical methods were used to identify significant correlations between water-quality variables and potential explanatory variables. Non-parametric statistics are robust techniques that generally are not affected by outliers and do not require that the data follow any particular distribution (Helsel and Hirsch, 2002). The significance level (p) generated from hypothesis testing for this report was compared to a threshold value (α) of 5 percent ($\alpha = 0.05$) to evaluate whether the relation was statistically significant ($p < \alpha$). Two different statistical tests were used because the set of potential explanatory factors included continuous and categorical variables. Relations between categorical variables (for example, groundwater age or land-use classes) and continuous variables were evaluated using the Wilcoxon rank-sum test. Correlations between continuous variables were evaluated using Spearman's method. Correlations between potential explanatory factors, between water-quality parameters, and between potential explanatory factors and water-quality constituents were tested for significance. Correlations on pesticides, THMs, and solvents were performed on the sum of their concentrations. For example, the sum of solvent concentrations for SCRV-13 is 1.27 µg/L [0.09 (perchloroethene) + 0.07 (carbon tetrachloride) + 0.12 (*trans*-1,2-dichloroethene) + 0.99 (trichloroethene)].

The wells selected for the statistical tests were USGS- and CDPH-grid wells and USGS-understanding wells, with the exception of well SCRVU-10. Well SCRVU-10 is a shallow monitoring well located on the coast that has been significantly affected by seawater and was excluded because samples collected from the well do not represent the primary aquifer system. Data from CDPH-other wells were not used in the understanding assessment because carbon isotope, tritium, dissolved oxygen, and some well construction data were not available. Correlations between explanatory factors and groundwater constituents were tested using data from the USGS- and CDPH-grid wells only, or the USGS- and CDPH-grid wells plus USGS-understanding wells. Because the USGS-understanding wells primarily represented groundwater in agricultural areas that were not randomly selected on a spatially distributed grid, the USGS-understanding wells were excluded from analysis of relations between water quality and areally distributed explanatory variables (land use and lateral position), to avoid areal-clustering bias. However, USGS-understanding wells were included in the analysis of relations between water quality and vertically distributed explanatory variables (depth, groundwater age, and oxidation-reduction characteristics).

Evaluation of Potential Explanatory Factors

Brief descriptions of potential explanatory factors, including land use, septic systems density, well depth, depth to the top-of-well perforations, groundwater age, and geochemical conditions, are included in this section. Data sources and methodology used for assigning values for potential explanatory factors are described in appendix D.

Land Use

In the SCRV study unit, land use is a combination of agricultural, urban, and natural; however, land use in the areas surrounding the study unit primarily is natural (fig. 4). Land use in the study unit was 37 percent agricultural, 23 percent urban, and 40 percent natural on the basis of Nagasaki and others (2007). Land use within a 500-m (1,640-ft) radius around each of the grid wells, on average, had a higher proportion of agricultural (42 percent) and urban (30 percent) land uses, and a lower proportion of natural land use (28 percent) (fig. 3), probably because wells tend to be located in areas of human settlement. In contrast, the land use within a 500-m radius around each of the CDPH wells, on average, had lower agricultural land use (33 percent), and even higher urban land use (37 percent) than the study unit as a whole. In general, the average agricultural and urban land use within the 500-m (1,640-ft) radius of grid wells is within 10 percent of the agricultural and urban land use of the study unit as a whole.

The predominant land use varies spatially across the study unit. For example in the Santa Clara River Valley East subbasin (easternmost part of the Santa Clara River Valley basin), the land use is predominantly natural, with the exception of the urban areas in and around the cities of Newhall and Santa Clarita (fig. 4). The remainder of the Santa Clara River Valley basin (excluding Santa Clara River Valley East subbasin) is predominantly agricultural, with distinct urban land use in and around the cities of Santa Paula, Ventura, and Oxnard, and natural land use around the Santa Clara River floodplain. The Simi Valley groundwater basin is predominantly urban.

Septic Systems

Septic system density within the 500-m (1,640-ft) radius around the USGS-grid and USGS-understanding wells ranged from 0 to 137 septic tanks per km^2, with a median of 5 septic tanks per km^2. Many of the wells in areas with septic system densities greater than median were located in the Oxnard subbasin or the Ventura River Valley, Ojai Valley, or upper Ojai Valley basins (table D1).

Depth

Depths of USGS- and CDPH-grid wells varied across the study unit. Well depths in grid wells ranged from 60 to 1,440 ft below land surface, with a median of 541 ft (fig. 7). Depth to top-of-perforations ranged from 15 to 800 ft, with a median of 403 ft. The maximum perforation length was 690 ft, with a median of 220 ft. These values represent a subset of the USGS- and CDPH-grid wells because the well depths and depth to top-of-perforations were not known for every well. Only the wells with construction information available were included in the analyses of the explanatory variables (table D2).

The median depth of USGS-understanding wells was deeper than the median depth for USGS- and CDPH-grid wells (fig. 7). The median depth for USGS-understanding wells was 750 ft. However, the median of the depth to top-of-perforations and perforation lengths were similar for grid and understanding wells.

Groundwater Age

Groundwater samples were assigned age classifications based on the tritium, helium-3, carbon-14, and helium-4 content of the samples (appendix D). Of the 53 grid and understanding wells sampled by USGS-Priority Basin Project, groundwater samples were classified as modern age in 17 wells, mixed age in 18 wells (evidence of modern and pre-modern groundwater in the same sample), and pre-modern age in 13 wells (table D3). Samples from five wells could not be classified because the age-tracer data were incomplete or did not meet all quality-assurance checks.

Wells with groundwater classified as pre-modern age generally were deeper than the wells with groundwater classified as modern or mixed age (fig. 8A, table 6). The depths to top-of-perforations of wells with groundwater classified as pre-modern age generally were deeper than the depths to top-of-perforations of wells with groundwater classified as modern or mixed age (fig. 8B, table 6).

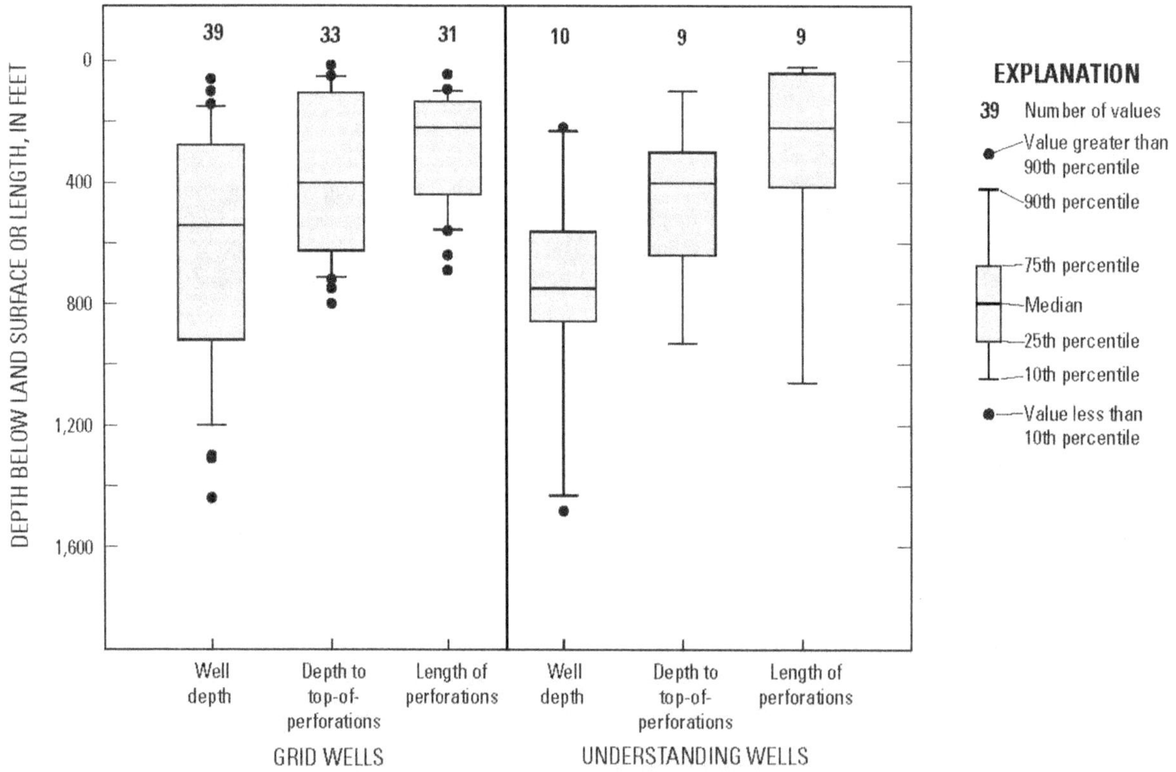

Figure 7. Well depths, depths to top-of-perforations, and perforation lengths, for USGS- and CDPH-grid and USGS-understanding wells, Santa Clara River Valley study unit, California GAMA Priority Basin Project.

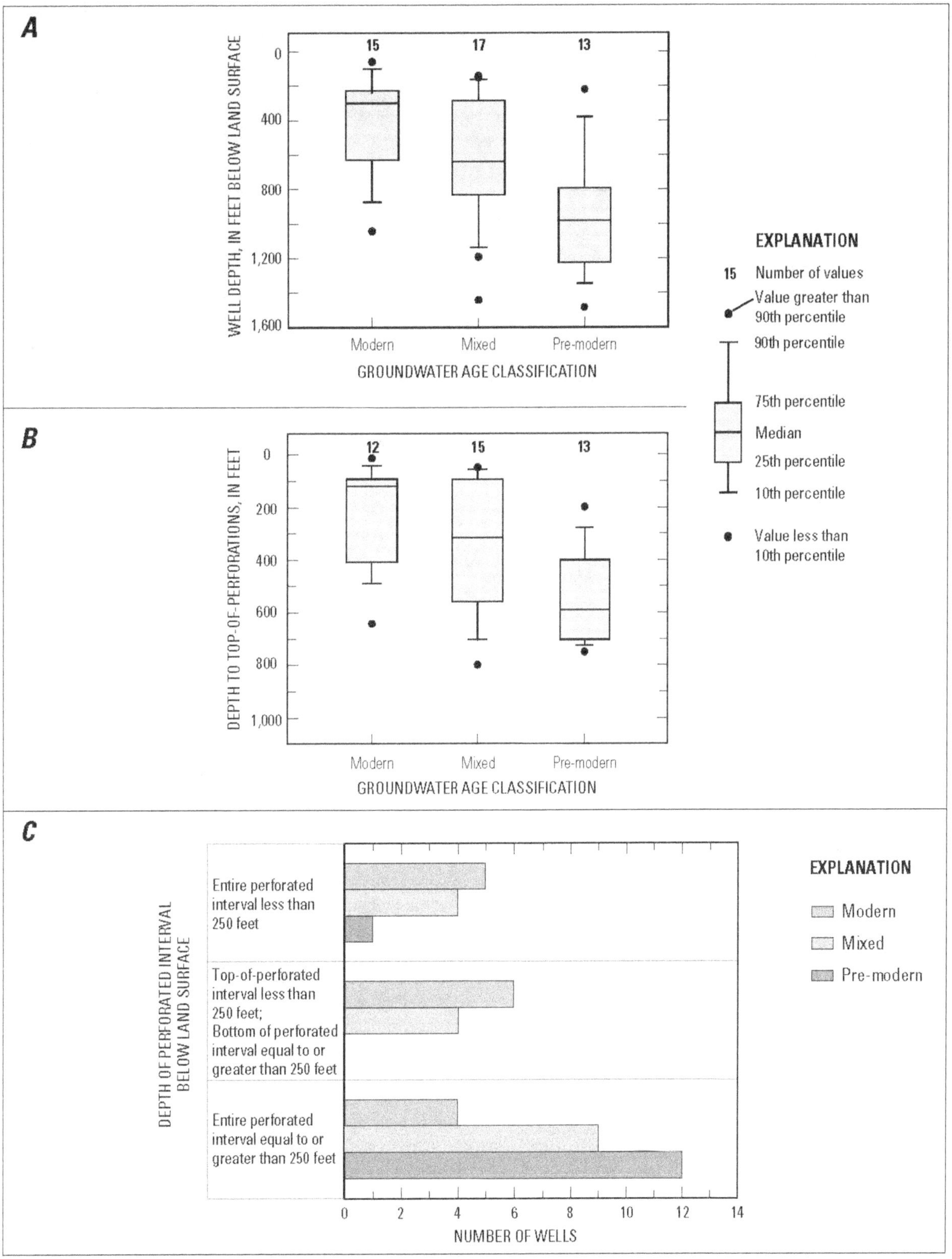

Figure 8. Relation of groundwater age classification to (*A*) well depth and (*B*) depth to top-of-perforations, and (*C*) numbers of wells with each groundwater age class in each of the three depth categories, Santa Clara River Valley study unit, California GAMA Priority Basin Project.

Table 6. Results of non-parametric (Wilcoxon) statistical tests for differences in values of potential explanatory factors and selected water-quality constituents between modern, mixed, and pre-modern groundwater age classes, Santa Clara River Valley study unit, California, GAMA Priority Basin Project.

[Wilcoxon rank sum tests; Z, test statistic for Wilcoxon test; Z statistic is shown when the difference between age classes is significant ($p < 0.05$); positive Z statistic (values of the selected explanatory factor or constituent of the first age class generally are greater than second age class); negative Z statistic (values of selected explanatory factor or constituent of first age class generally are smaller than the second age class); GAMA, Groundwater Ambient Monitoring and Assessment Program; ns, no significant difference]

	Groundwater age class comparisons		
	Modern compared with mixed	Mixed compared with pre-modern	Modern compared with pre-modern
	Z	Z	Z
Potential explanatory factors			
Depth to top-of-perforations	ns	−2.051	−3.265
Well depth	ns	−2.470	−3.250
Dissolved oxygen	ns	ns	2.835
pH	ns	−2.042	−2.440
Selected water-quality constituents			
Total dissolved solids (TDS)	ns	ns	ns
Sulfate	ns	ns	ns
Manganese	ns	ns	ns
Iron	ns	ns	−2.244
Nitrate as nitrogen	ns	2.806	3.993
Vanadium	ns	ns	2.298
Arsenic	ns	ns	ns
Boron	ns	ns	ns
Gross alpha radioactivity	ns	ns	ns
Perchlorate	ns	ns	ns
Sum of pesticide concentrations	ns	ns	ns
Sum of solvent concentrations	ns	ns	ns
Total trihalomethane concentrations	ns	ns	ns

Nearly all wells with perforation intervals entirely between land surface and 250 ft below land surface (9 of 10 wells) had groundwater ages that were modern or mixed (fig. 8C). Likewise, many wells with perforation intervals entirely at depths equal to or greater than 250 ft (12 of 25 wells) had groundwater ages that were pre-modern. Wells with perforation intervals that crossed the 250 ft depth had groundwater ages that were modern or mixed. The 250 ft depth criterion was arbitrarily selected; groundwater age generally was younger, dissolved oxygen and nitrate concentrations generally were higher, and more pesticide and volatile organic compounds were detected at depths less than 250 ft than at depths greater than 250 ft.

The presence of one pre-modern age sample from a well less than 250 ft deep and four modern age samples from wells greater than or equal to 250 ft deep (fig. 8C) indicate that there are local variations in the general groundwater age-depth relations. These variations may indicate the position of the well relative to regional recharge and discharge areas. For example, three wells with the top-of-perforation greater than 250 ft deep and with modern ages are located adjacent to the Santa Clara River (wells SCRV-15, SCRV-16, and SCRV-27, fig. A1A) (Montrella and Belitz, 2009); and another well (SCRV-24) is located in close proximity to a recharge facility and a golf course; the groundwater system may have relatively higher recharge rates in the vicinity of these wells than in other parts of the system, allowing modern age groundwater to infiltrate to greater depths than generally is the case elsewhere in the study unit.

Geochemical Conditions

An abridged classification of oxidation-reduction (redox) conditions adapted from the framework presented by McMahon and Chapelle (2008) is given in appendix D for the USGS-grid and USGS-understanding wells in the SCRV (table D4). In general, groundwater in the Santa Clara River Valley study unit primarily is oxic in shallow wells (fig. 9*A*) and in wells located in the upgradient basins (that is, Ojai Valley and Ventura River Valley basins, and Santa Clara River Valley East subbasin) but becomes more reducing or anoxic with depth and near the western (distal or downgradient) end of the study unit. Redox conditions for about one-half of groundwater samples (25 of 52 wells with complete data for redox characterization) are anoxic, ranging from nitrate-reducing to sulfate-reducing conditions (table D4). Redox conditions in the Santa Clara River Valley East, Piru, and Fillmore subbasins, and the Ventura River Valley, Ojai Valley, Upper Ojai Valley, and Arroyo Santa Rosa Valley basins primarily were oxic. Redox conditions in the Santa Paula, Mound, and Oxnard subbasins, and the Las Posas Valley basin primarily were anoxic. Groundwater samples from the Simi Valley and Pleasant Valley basins indicated oxic or anoxic conditions in different parts of these basins.

Dissolved-oxygen (DO) concentrations were used as the primary indicator variable for redox conditions for comparison with concentrations of water-quality constituents in this report. DO was used because DO data were available for 52 of the 53 USGS-GAMA sampled wells; more wells than other redox indicators. DO concentrations were not available for CDPH-grid wells that were used to supplement the USGS-grid well data.

DO concentrations were significantly greater for samples from wells with modern or mixed groundwater ages than for wells with pre-modern groundwater ages (table 6, fig. 9). DO concentrations were not significantly different between samples from wells with modern groundwater age and wells with mixed groundwater age. DO concentrations ranged from less than 0.2 to 8 mg/L across the study unit (table D4).

Wilcoxon tests indicate pH values were significantly lower for wells with mixed and modern groundwater ages than for wells with pre-modern groundwater ages (table 6). Values of pH were not significantly different between the wells with modern groundwater ages and the wells with mixed groundwater ages. Values of pH ranged from 6.8 to 7.8 in the study unit (table D4).

Correlations Between Explanatory Variables

Significant correlations between explanatory variables are important to identify because apparent correlations between an explanatory variable and a water-quality constituent could indicate relations between two explanatory variables and not between an explanatory variable and a water-quality constituent. Significant correlations between explanatory variables are shown in table 7 and are discussed along with the relations between groundwater age and depth. Implications of cross-correlations between explanatory variables are discussed later in the report as part of analysis of factors affecting individual water-quality constituents.

DO concentrations were negatively correlated with depth to top-of-perforations and well depth (table 7). These correlations indicate that DO concentrations decrease with increasing depth to top-of-perforations. These correlations were expected based on previous studies in other aquifers (McMahon and Chapelle, 2008). The wells in figure 9*A* represent all the wells with construction information (depth to top-of-perforations), DO, and groundwater age data in the study unit. The highest DO concentrations are from wells with shallow perforations, which typically have modern or mixed ages. DO concentrations were less than 1.0 mg/L for groundwater classified as pre-modern, with the exception of one well (SCRV-14, 1.8 mg/L, table D4). DO concentrations were more variable for groundwater ages classified as modern or mixed than for groundwater ages classified as pre-modern (fig. 9*A*).

The pH was positively correlated with well depth and depth to top-of-perforations (table 7), indicating increasing pH with increasing depth (fig. 9*B*). These correlations were expected based on previous studies in other aquifers in California (Jurgens and others, 2010) and reflect dissolution of primary minerals, causing the pH of groundwater to increase with depth and continued contact of groundwater with aquifer materials. Izbicki and Martin (1997) used geochemical modeling to conclude that weathering of silicate minerals, which causes increased pH, was an important process along groundwater flow paths in the Las Posas basin. The pH was negatively correlated with urban land use (table 7), indicating increasing pH with decreasing percentage of urban land use. This correlation was to be expected because most wells in the SCRV with urban land use less than 20 percent were in predominantly agricultural areas (fig. 3); and wells in the SCRV with predominantly agricultural land use tend to be relatively deep (on the basis of the positive correlation of agricultural land use and well depth); therefore, wells with lower percentage of urban land use are likely to have a higher pH (on the basis of the positive correlation of pH and well depth; for example, well SCRV-20).

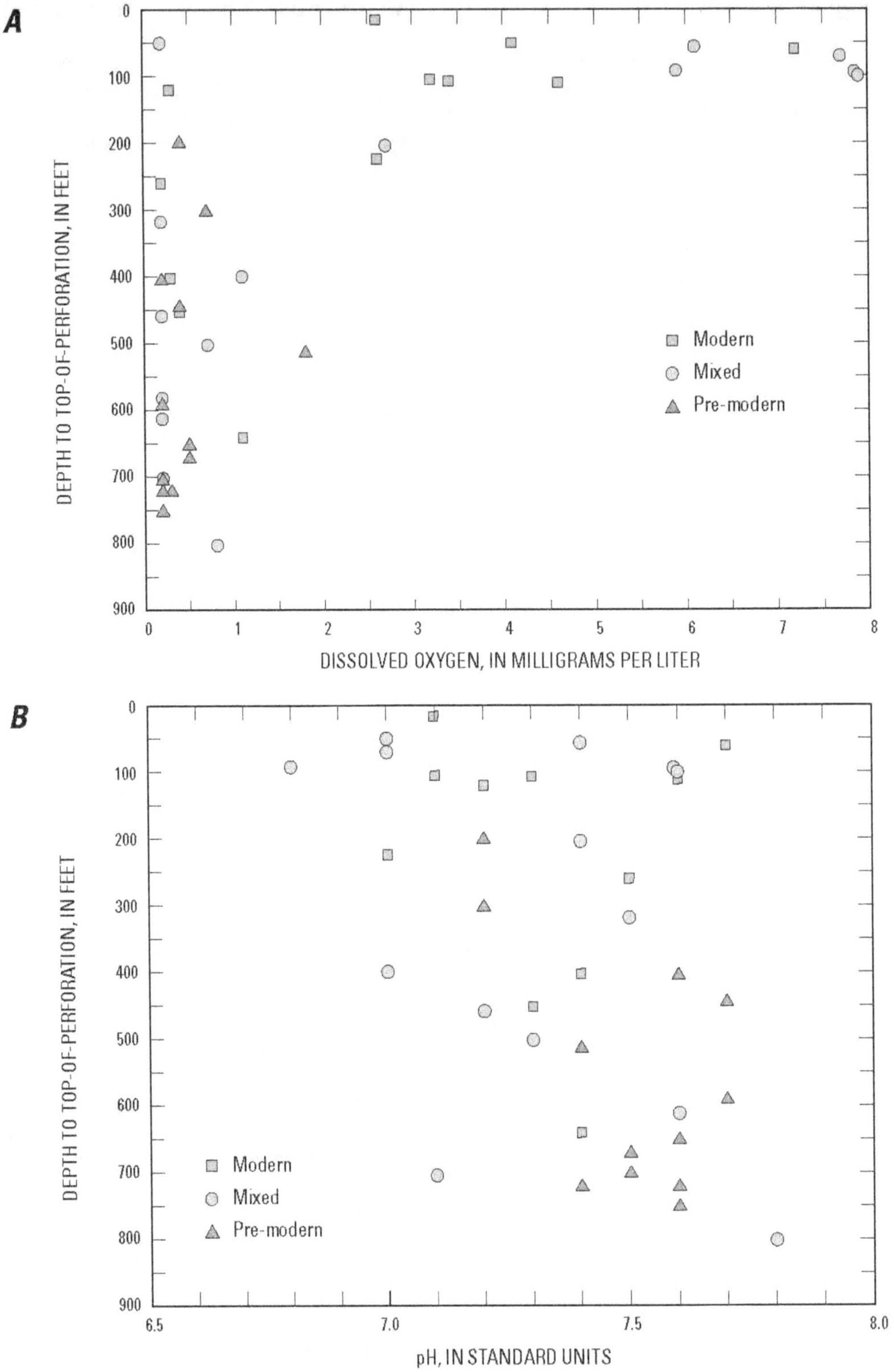

Figure 9. (*A*) Relation of dissolved-oxygen concentration (an oxidation-reduction condition indicator), depth to top-of-perforations, and groundwater age class, and (*B*) relation of pH, depth to top-of-perforations, and groundwater age class, Santa Clara River Valley study unit, California GAMA Priority Basin Project.

The number of septic tanks or cesspools was negatively correlated with percentage of natural land use (table 7). This correlation was expected because fewer septic tanks or cesspools would be expected because fewer people live in natural land-use areas than in areas of agricultural or urban land use.

Well depth (and depth to top-of-perforations) was positively correlated with percentage of agricultural land use (table 7). These correlations indicate that wells are deeper in predominantly agricultural areas in this study unit than in areas that are predominantly urban or natural land use. In particular, seven wells with more than 80 percent agricultural land use were greater than 800 ft deep, putting these wells in the deepest 25 percent of wells included in the study. Intensive use of groundwater for agriculture has lowered the water table; thus, wells need to be deeper. Correspondingly, well depth (and depth to top-of-perforations) was negatively correlated with percentage of natural land use (table 7). These correlations indicate that wells are shallower in predominantly natural areas.

Table 7. Results of non-parametric (Spearman's *rho* method) analysis of correlations between selected potential explanatory factors, Santa Clara River Valley study unit, California, GAMA Priority Basin Project.

[GAMA, Groundwater Ambient Monitoring and Assessment Program; Spearman's *rho* test used and *rho* values shown when correlations between selected explanatory factors are significant ($p<0.05$). Land-use percentages and number of septic tanks or cesspools within a circle with a radius of 500 meters centered around each well. <, less than; ns, not significant]

Wells included in analysis	Explanatory factor	Dissolved-oxygen concentration	pH	Number of septic tanks or cesspools	Depth to top-of-perforation	Well depth
Grid wells	Percentage of urban land use	ns	-0.336	ns	ns	ns
	Percentage of agricultural land use	ns	ns	ns	0.518	0.516
	Percentage of natural land use	ns	ns	−0.360	−0.501	−0.467
	Number of septic tanks or cesspools	ns	ns		ns	ns
Grid and understanding wells	Depth to top-of-perforation	-0.614	0.329			
	Well depth	-0.585	0.345			

Status and Understanding of Water Quality

The *status assessment* was designed to identify the constituents or classes of constituents most likely to be of water-quality concern because of their high relative-concentrations or their prevalence. The assessment applies only to constituents with regulatory (MCL and AL) or non-regulatory (HAL, RSD5-US, or NL) human-health benchmarks or aesthetic benchmarks (SMCL) established by the USEPA or the CDPH (California Department of Public Health, 2008a; U.S. Environmental Protection Agency, 2008a, 2008b). The spatially distributed, randomized approach to well selection and data analysis yields a view of groundwater quality in which all areas of the primary aquifer system are weighted equally.

The *understanding assessment* was designed to help answer the question of why specific constituents are, or are not, observed in groundwater. The understanding assessment addresses a subset of the constituents discussed in the status assessment and is based on statistical correlations between water quality and a finite set of potential explanatory factors. This assessment may improve our understanding of how human and natural sources of contaminants affect groundwater quality in the SCRV; however, it was not designed to identify specific sources of constituents to specific wells.

Human-health benchmarks are established for about one-half of the organic and special interest constituents (21 of 39) that were detected in USGS-grid wells (table 8).

Table 8. Number of constituents analyzed for, and number of constituents detected in, U.S. Geological Survey (USGS) grid wells listed by human-health-based or aesthetic benchmark type and constituent type, Santa Clara River Valley study unit, California, GAMA Priority Basin Project.

[Regulatory health-based benchmarks include U.S. Environmental Protection Agency (USEPA) and California Department of Public Health (CDPH) maximum contaminant levels. Nonregulatory health-based benchmarks include USEPA lifetime health advisory levels, risk specific dose level at 10^{-5} lifetime cancer risk, and CDPH notification level. GAMA, Groundwater Ambient Monitoring and Assessment Program; HHB, human-health benchmark; NWQL, U.S. Geological Survey (USGS) National Water Quality Laboratory; SMCL, secondary maximum contaminant level (aesthetic benchmark); VOC, volatile organic compound]

Benchmark type	Number of constituents analyzed	Number of constituents detected
VOCs plus gasoline oxygenates		
Regulatory – HHB	33	13
Nonregulatory – HHB	26	2
No benchmark	29	2
Total:	88	17
Pesticides and degradates (NWQL Schedule 2003)		
Regulatory – HHB	3	2
Nonregulatory – HHB	17	3
No benchmark	43	8
Total:	63	13
Polar pesticides and degradates (NWQL Schedule 2060)		
Regulatory – HHB	10	0
Nonregulatory – HHB	9	0
No benchmark	36	0
Total:	55	0
Potential wastewater indicator compounds		
Regulatory – HHB	2	0
Nonregulatory– HHB	2	0
No benchmark	50	8
Total:	54	8
Special interest (perchlorate)		
Regulatory – HHB	1	1
Total:	1	1
Sum organic and special interest constituents		
Regulatory – HHB	48	16
Nonregulatory – HHB	53	5
No benchmark	159	18
Total:	260	39
Sum inorganic and radioactive constituents		
Regulatory – HHB	22	21
Nonregulatory – HHB	8	8
Nonregulatory – SMCL	6	6
No benchmark	17	17
Total:	53	52

Seventeen VOCs, including gasoline oxygenates, were detected; human-health benchmarks are established for all but two VOCs. Thirteen pesticides were detected; human-health benchmarks are established for 5 of the 13 pesticides. Five of the pesticides with no benchmarks (de-ethylatrazine, desulfinylfipronil, 3,4-dichloroanaline, fipronil sulfide, and fipronil sulfone) are degradates; human-health benchmarks are established for two of the three parent compounds (atrazine and diuron) of these degradates. Human-health benchmarks have not been established for the parent compound (fipronil) of the remaining degradates that were detected. Thus, the organic constituents that are regulated include most of these constituents that were detected in groundwater in the SCRV. Human-health benchmarks have not been established for the eight potential wastewater indicator compounds that were detected (table 8). Four of the wastewater compounds were detected at concentrations greater than the compounds respective reporting levels (Montrella and Belitz, 2009).

In contrast to organic constituents, inorganic constituents were nearly always detected (52 of 53, table 8) in USGS- and CDPH-grid wells. Human-health or aesthetic benchmarks have been established for 36 of the 53 inorganic constituents. Most of the constituents without benchmarks are major or minor ions that are naturally present in nearly all groundwater.

The maximum relative-concentration for each constituent with a water-quality benchmark in grid wells is shown in figure 10. Eight inorganic constituents (including radioactive constituents) were detected at high relative-concentrations in one or more grid wells; boron was also detected at high relative-concentrations, but only in a few CDPH-other wells. In contrast, none of the organic and special-interest constituents were detected at high relative-concentrations in grid wells, and only four organic and special-interest constituents were detected at moderate relative-concentrations. Aquifer-scale proportions were calculated for each constituent detected at high or moderate relative-concentrations and for each organic and special-interest constituent detected in more than 10 percent of the grid wells. Spatially weighted, high aquifer-scale proportions were within the 90 percent confidence intervals for their respective grid-based aquifer high proportions (16 of 16 constituents; see table 9), providing evidence that the grid-based approach yields statistically equivalent results to the spatially weighted approach.

Inorganic Constituents

Inorganic constituents generally occur naturally in groundwater, although their concentrations may be influenced by humans as well as natural factors. Inorganic constituents with human-health benchmarks, as a group (trace elements, fluoride, radioactive constituents, and nutrients), had high aquifer-scale proportions in 21 percent, moderate in 30 percent, and low or not detected in 49 percent of the primary aquifer system (table 10). Inorganic constituents with aesthetic benchmarks, as a group, had high aquifer-scale proportions in 54 percent, moderate in 41 percent, and low in 4.3 percent of the primary aquifer system.

Trace Elements and Minor Ions

Trace elements and minor ions with human-health benchmarks, as a class, had high relative-concentrations for one or more constituents in 4.4 percent, moderate in 38 percent, and low or not detected in 58 percent of the primary aquifer system (table 10). Percentages may not add to 100 as a result of rounding. Relative-concentrations of arsenic, vanadium, and boron were high in at least one grid well (table 9).

Arsenic was detected at high relative-concentrations in 2.3 percent of the primary aquifer system and was low or not detected in 98 percent (table 9, fig. 11). Arsenic was detected at a high relative-concentration of in one well in the foothills of the San Gabriel Mountains in the Santa Clara River Valley East subbasin (fig. 12*A*).

Vanadium was detected at high relative-concentrations in 3.4 percent of the primary aquifer system and, like arsenic, was not detected at moderate relative-concentrations (table 9, fig. 11). Vanadium was detected at a high relative-concentration in one well in the Arroyo Santa Rosa Valley groundwater basin (fig. 12*B*).

Boron was detected at high relative-concentrations in 3.2 percent of the primary aquifer system, as determined by using the spatially weighted approach. Boron was detected at high relative-concentrations in the Fillmore and Oxnard subbasins and in the Simi Valley and Pleasant Valley basins (fig. 12*C*). Boron was detected at moderate relative-concentrations mostly in the Santa Clara River Valley, Pleasant Valley, and Simi Valley basins (fig. 12*C*).

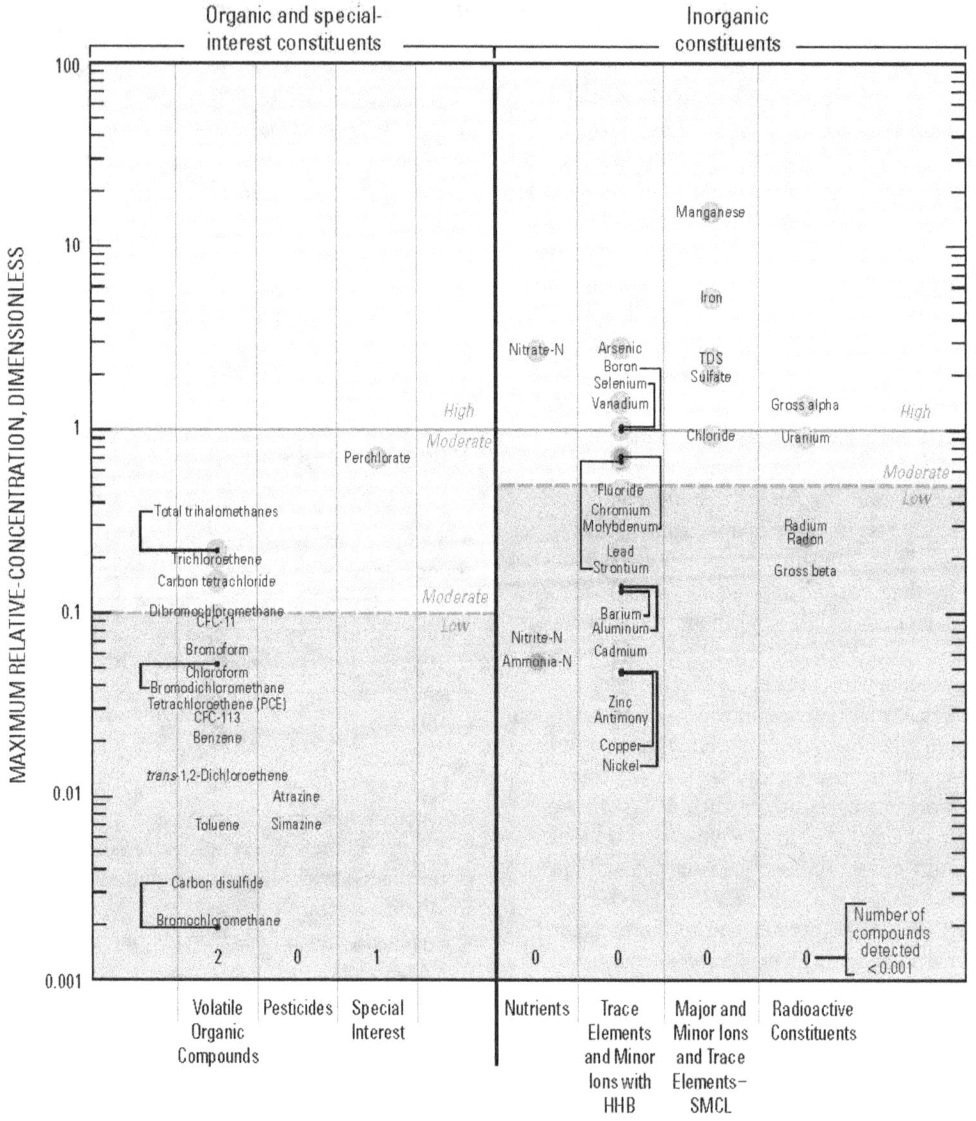

EXPLANATION

Silver — **Constituents with analyses in ≥24 grid wells and wells are spatially representative**
Name and center of symbol is location of data unless indicated by following location line

Strontium — **Constituents with analyses in <24 grid wells and wells are spatially representative**
Name and center of symbol is location of data unless indicated by following location line

Abbreviations: N, Nitrogen; TDS, Total Dissolved Solids; HHB, Human-Health Benchmark

Figure 10. Maximum relative-concentration in USGS-grid and CDPH-grid wells for constituents detected, grouped by type of constituent, Santa Clara River Valley study unit, California GAMA Priority Basin Project.

Table 9. Raw-detection frequency and aquifer-scale proportions calculated by using spatially weighted and grid-based approaches for constituents detected at concentrations greater than water-quality benchmarks during the most recent 3 years of available data (November 1, 2003–October 31, 2006) from the California Department of Public Health (CDPH) database, detected at high or moderate relative-concentrations in USGS-grid wells sampled April–June 2007, or organic constitutents detected in greater than 10 percent of USGS-grid wells sampled, Santa Clara River Valley study unit, California, GAMA Priority Basin Project.

[High, concentrations greater than water-quality benchmark; moderate, concentrations less than or equal to the water-quality benchmark, but greater than 0.1 of the benchmark (for organic and special interest constituents) or greater than 0.5 of the benchmark (for inorganic constituents); low, concentrations less than or equal to 0.1 of benchmark (for organic and special interest constituents) or less than or equal to 0.5 of the benchmark (inorganic constituents); GAMA, Groundwater Ambient Monitoring and Assessment Program]

Constituent	Raw detection frequency [1]			Spatially weighted aquifer-scale proportion [1] (Cell declustered)			Grid-based aquifer-scale proportion			90-percent confidence interval for grid-based high aquifer-scale proportion [2]	
	Number of wells	Number of high wells	High aquifer-scale proportion (percent)	Number of cells	Moderate values (percent)	High values (percent)	Number of cells	Moderate values (percent)	High values (percent)	Lower limit (percent)	Upper limit (percent)
Major and minor elements											
Total dissolved solids (TDS)	166	47	28	45	54	33	46	56	35	24	47
Sulfate	163	25	15	44	32	20	45	27	22	13	34
Manganese	168	48	29	44	3.3	32	45	2.2	38	27	50
Iron	165	30	18	44	6.0	24	45	4.4	22	13	34
Chloride	163	0	0.0	44	2.3	0.0	45	2.2	0.0	0.0	2.9
Fluoride	163	0	0.0	44	0.3	0.0	45	0.0	0.0	0.0	2.9
Nutrients											
Nitrate plus nitrite, as nitrogen	193	14	7.3	45	8.9	11	46	8.7	15	8.2	25
Trace elements											
Vanadium	96	3	3.1	33	0.0	3.0	29	0.0	3.4	0.6	13
Arsenic	159	1	0.6	43	0.0	1.2	44	0.0	2.3	0.4	8.5
Boron	117	5	4.3	34	50	3.2	32	45	0.0	0.0	4.0
Selenium	159	0	0.0	43	5.8	0.0	44	6.8	0.0	0.0	3.0
Aluminum	159	0	0.0	44	0.4	0.0	45	0.0	0.0	0.0	2.9
Lead	123	0	0.0	38	1.3	0.0	36	0.0	0.0	0.0	3.7
Radioactive constituents											
Gross-alpha radioactivity, 72-hr count	148	6	4.1	39	11	7.3	29	6.9	14	5.9	27
Uranium	91	0	0.0	39	13	0.0	35	20	0.0	0.0	3.8
Trihalomethanes (disinfection by products) [3]											
Total trihalomethanes	128	0	0.0	45	0.7	0.0	42	2.4	0.0	0.0	4.4
Chloroform	128	0	0.0	45	0.0	0.0	42	0.0	0.0	0.0	4.4

Table 9. Raw-detection frequency and aquifer-scale proportions calculated by using spatially weighted and grid-based approaches for constituents detected at concentrations greater than water-quality benchmarks during the most recent 3 years of available data (November 1, 2003–October 31, 2006) from the California Department of Public Health (CDPH) database, detected at high or moderate relative-concentrations in USGS-grid wells sampled April–June 2007, or organic constitutents detected in greater than 10 percent of USGS-grid wells sampled, Santa Clara River Valley study unit, California, GAMA Priority Basin Project. —Continued.

[High, concentrations greater than water-quality benchmark; moderate, concentrations less than or equal to the water-quality benchmark, but greater than 0.1 of the benchmark (for organic and special interest constituents) or greater than 0.5 of the benchmark (for inorganic constituents); low, concentrations less than or equal to 0.1 of benchmark (for organic and special interest constituents) or less than or equal to 0.5 of the benchmark (inorganic constituents); GAMA, Groundwater Ambient Monitoring and Assessment Program]

Constituent	Raw detection frequency [1]			Spatially weighted aquifer-scale proportion [1] (Cell declustered)			Grid-based aquifer-scale proportion			90-percent confidence interval for grid-based high aquifer-scale proportion [2]	
	Number of wells	Number of high wells	High aquifer-scale proportion (percent)	Number of cells	Moderate values (percent)	High values (percent)	Number of cells	Moderate values (percent)	High values (percent)	Lower limit (percent)	Upper limit (percent)
Solvents											
Carbon tetrachloride (tetrachloromethane)	132	0	0.0	45	0.7	0.0	42	2.4	0.0	0.0	4.4
Trichloroethene (TCE)	132	0	0.0	45	0.7	0.0	42	2.4	0.0	0.0	4.4
Pesticides											
Simazine	121	0	0.0	45	0.0	0.0	42	0.0	0.0	0.0	4.4
Atrazine	121	0	0.0	45	0.0	0.0	42	0.0	0.0	0.0	4.4
Special interest											
Perchlorate	70	0	0.0	45	10.7	0.0	43	12	0.0	0.0	4.4

[1] Based on most recent CDPH analysis during November 1, 2003–October 31, 2006 (most recent data available at time of analysis) combined with grid-based data.

[2] Based on the Jeffreys interval for the binomial distribution (Brown and others, 2001).

[3] The MCL-US (U.S. Environmental Protection Agency maximum contaminant level) threshold for trihalomethanes is the sum of chloroform, bromoform, bromodichloromethane, and dibromochloromethane.

Table 10. Aquifer-scale proportions calculated for constituent classes, Santa Clara River Valley study unit, California, GAMA Priority Basin Project.

[Aquifer-scale proportions are given in percentage of area of the primary aquifer system. GAMA, Groundwater Ambient Monitoring and Assessment Program; THM, trihalomethane; TDS, total dissolved solids; SO_4, sulfate; Cl, chloride; VOC, volatile organic compound; SMCL, secondary maximum contaminant level. All values greater than 10 percent are rounded to the nearest 1 percent; values less than 10 percent are rounded to the nearest 0.1 percent; values may not add up to 100 percent because of rounding]

Constituent	Aquifer-scale proportion (percent)		
	High values	Moderate values	Low values or not detected
Inorganics with human-health benchmarks			
Trace elements and minor ions	4.4	38	58
Radioactive constituents	14	11	75
Nutrients	15	8.7	76
Any inorganic with human-health benchmarks	21	30	49
Inorganics with aesthetic benchmarks (SMCL)			
Major and minor ions (TDS, SO_4, Cl)	35	56	8.7
Manganese and (or) iron	44	0.0	56
Any inorganic constituent with an SMCL	54	41	4.3
Organics with human-health benchmarks			
Pesticides	0.0	0.0	100
Total THMs	0.0	2.4	98
Solvents	0.0	2.4	98
Other VOCs	0.0	0.0	100
Any organic constituent	0.0	4.8	95
Constituents of special interest			
Perchlorate	0.0	12	88

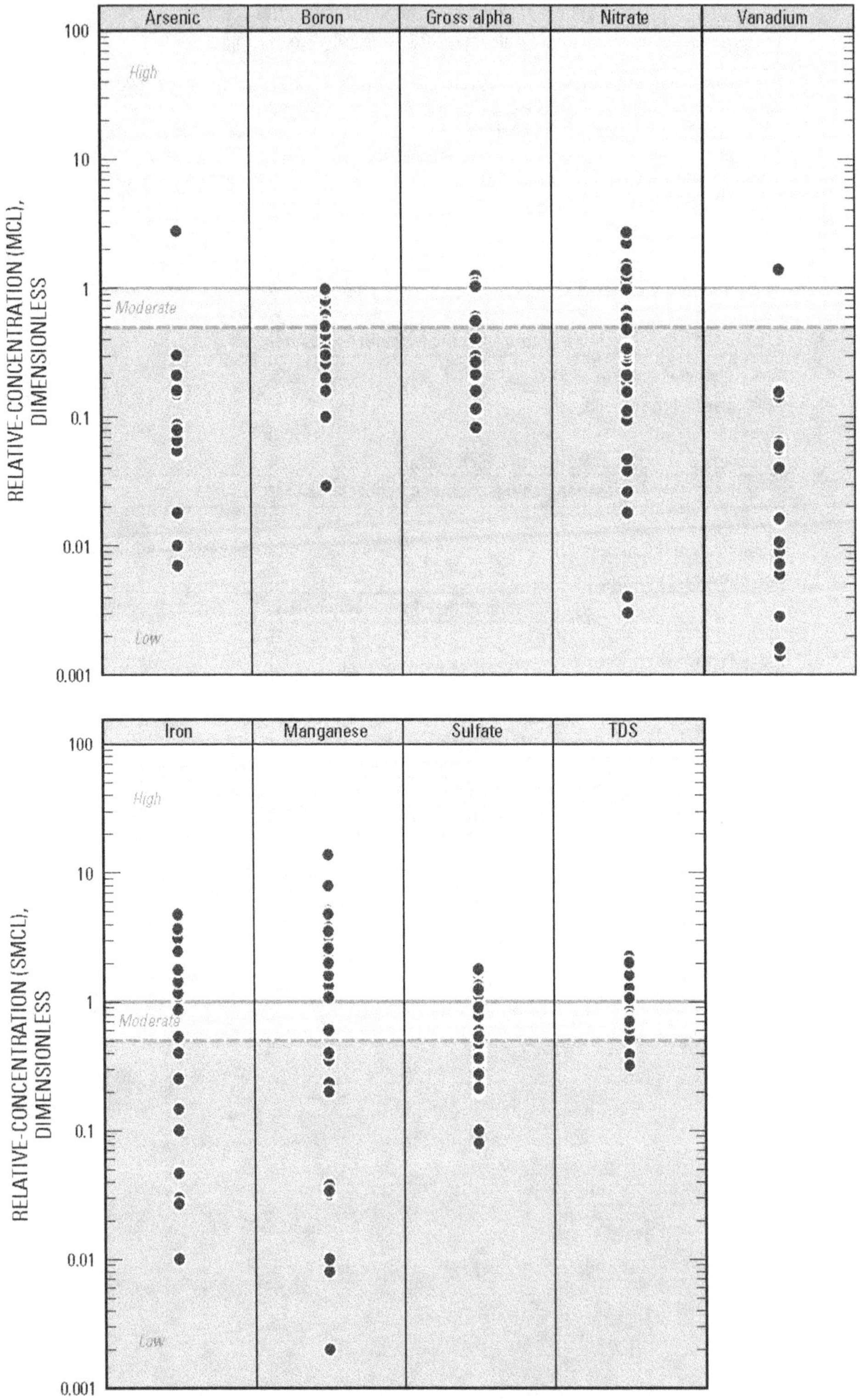

Figure 11. Relative-concentrations of inorganic constituents with human-health-based or aesthetic benchmarks categorized as high, medium, and low in USGS- and CDPH-grid wells, Santa Clara River Valley study unit, California GAMA Priority Basin Project.

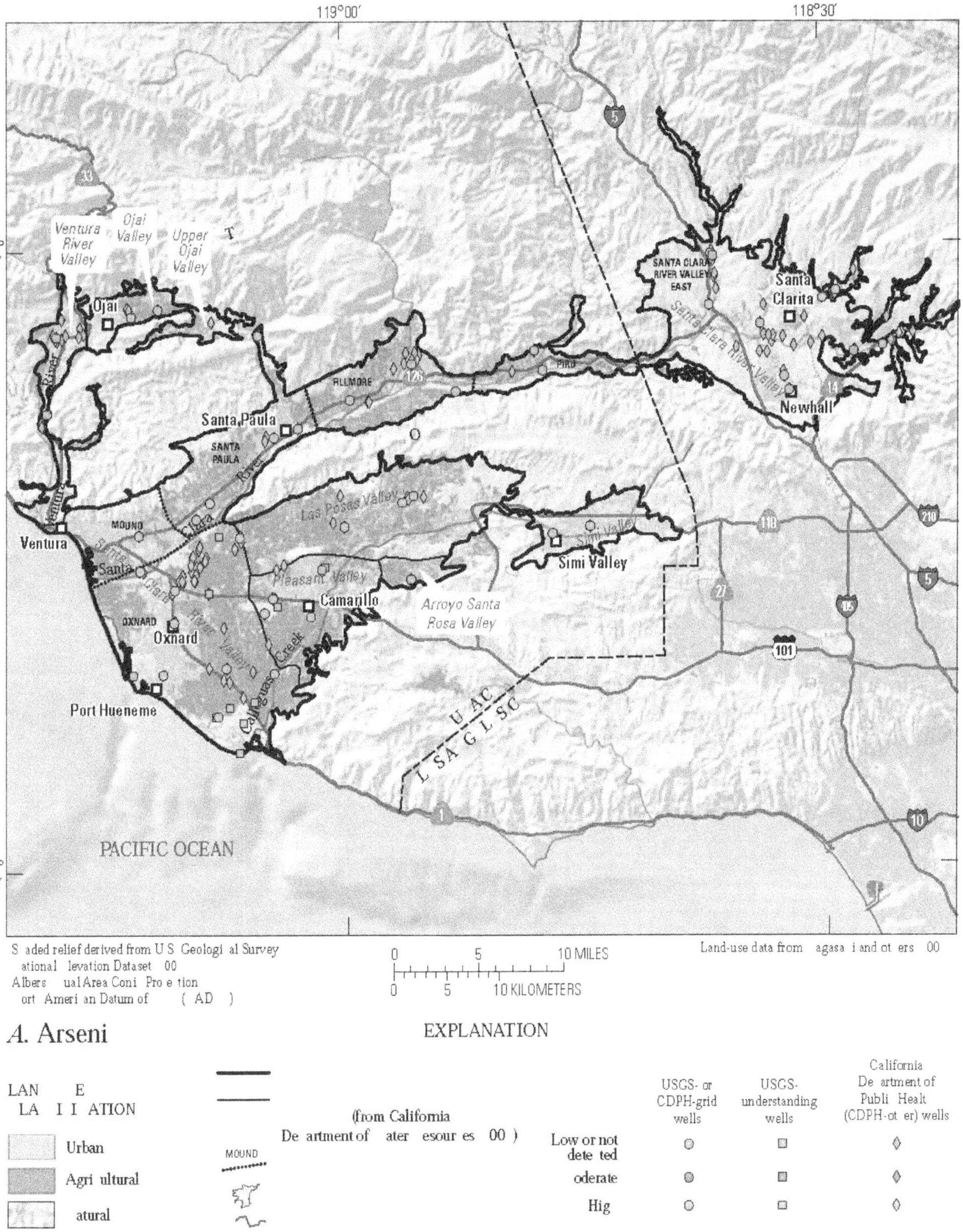

Figure 12. Relative-concentrations of selected inorganic constituents with human-health-based and aesthetic benchmarks in USGS-grid and CDPH-grid wells and CDPH-other wells in the Santa Clara River Valley study unit, California GAMA Priority Basin Project: (*A*) arsenic, (*B*) vanadium, (*C*) boron, (*D*) gross alpha radioactivity, (*E*) nitrate, (*F*) total dissolved solids, (*G*) sulfate, (*H*) manganese, and (*I*) iron.

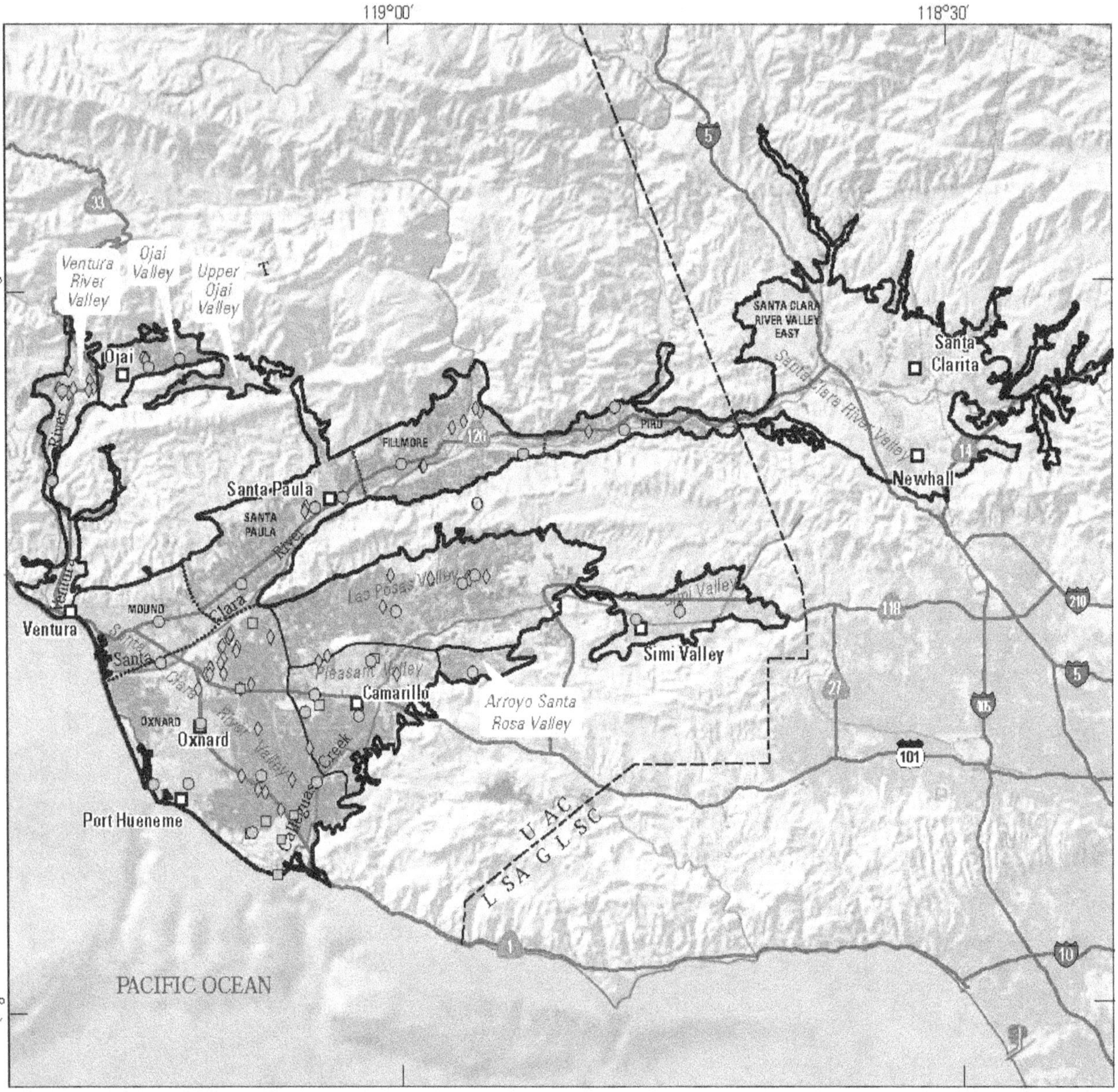

S aded relief derived from U S Geologi al Survey ational levation Dataset 00
Albers ual Area Coni Pro e tion ort Ameri an Datum of (AD)

0 5 10 MILES
0 5 10 KILOMETERS

Land-use data from a aga i and ot ers 00

B. anadium

EXPLANATION

LAN E
LA I I ATION

Urban
Agri ultural
atural

(from California De artment of ater esour es 00)

MOUND

	USGS- or CDPH-grid wells	USGS-understanding wells	California De artment of Publi Healt (CDPH-ot er) wells
Low or not dete ted	○	□	◊
oderate	○	□	◊
Hig	○	□	◊

Figure 12.—Continued

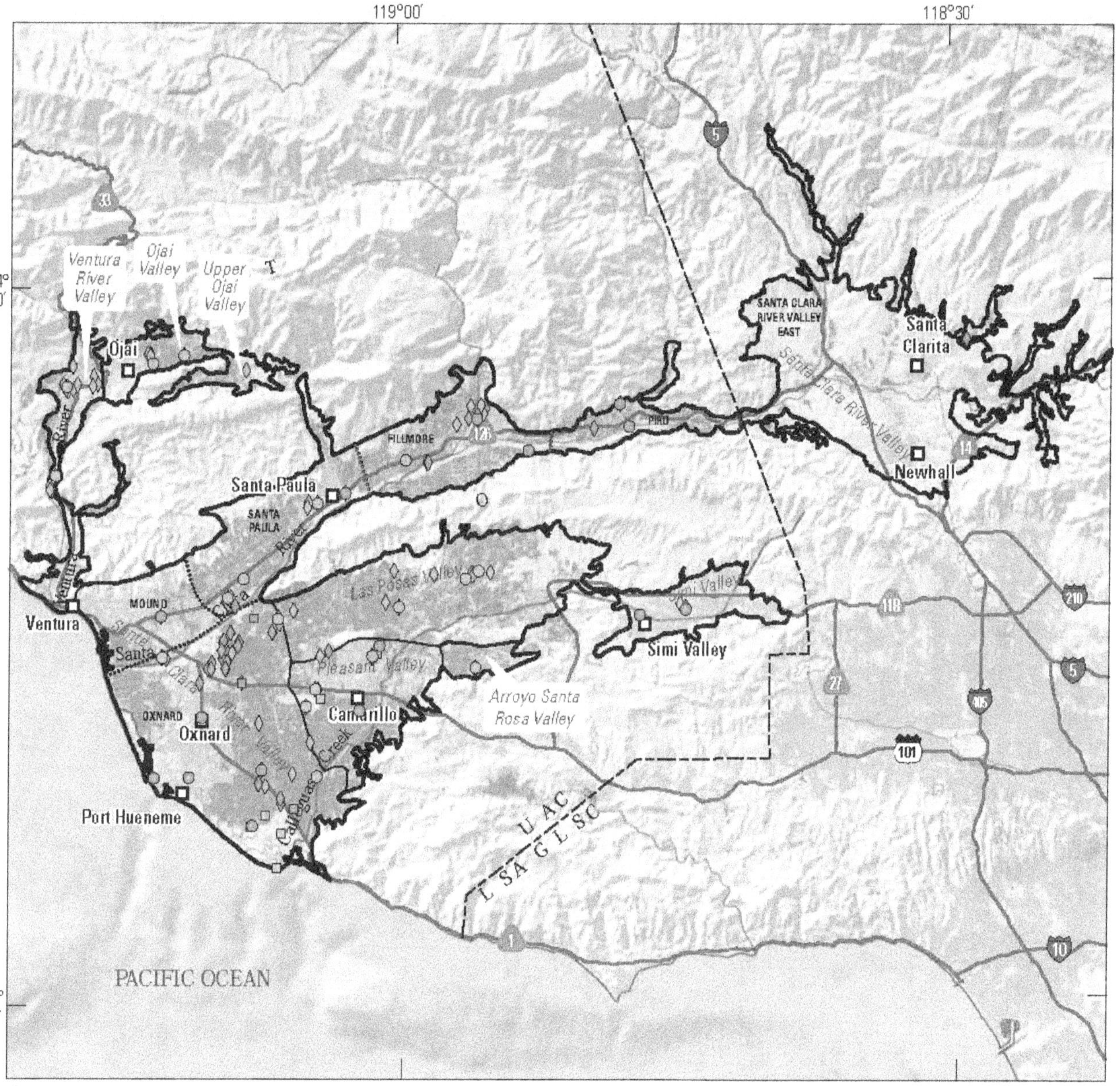

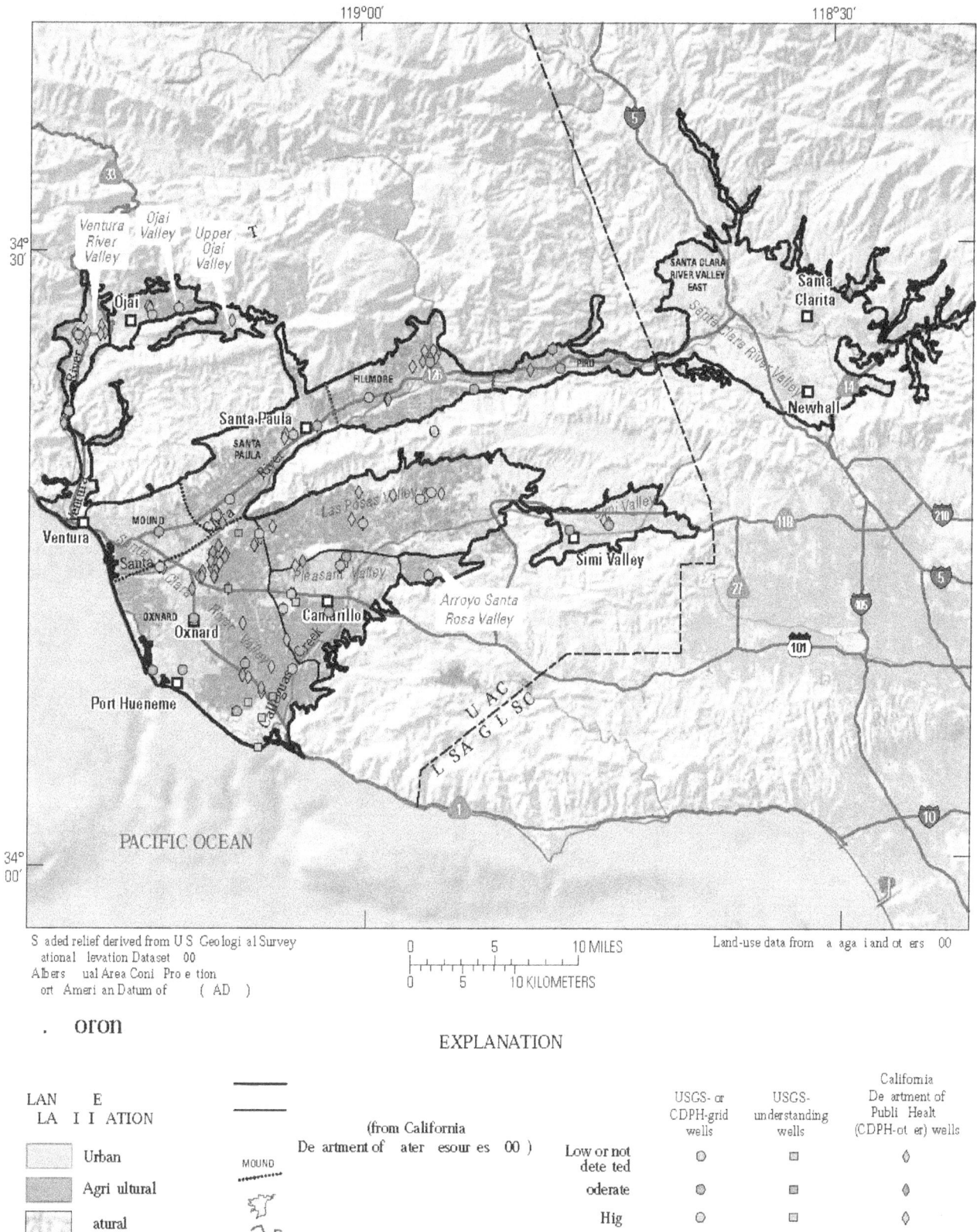

S aded relief derived from U S Geologi al Survey ational levation Dataset 00
Albers ual Area Coni Pro e tion
ort Ameri an Datum of (AD)

0 5 10 MILES
0 5 10 KILOMETERS

Land-use data from a aga i and ot ers 00

. oron

EXPLANATION

LAN E
LA I I ATION

Urban

Agri ultural

atural

(from California
De artment of ater esour es 00)

MOUND

	USGS- or CDPH-grid wells	USGS-understanding wells	California De artment of Publi Healt (CDPH-ot er) wells
Low or not dete ted	○	□	◊
oderate	○	□	◊
Hig	○	□	◊

Figure 12.—Continued

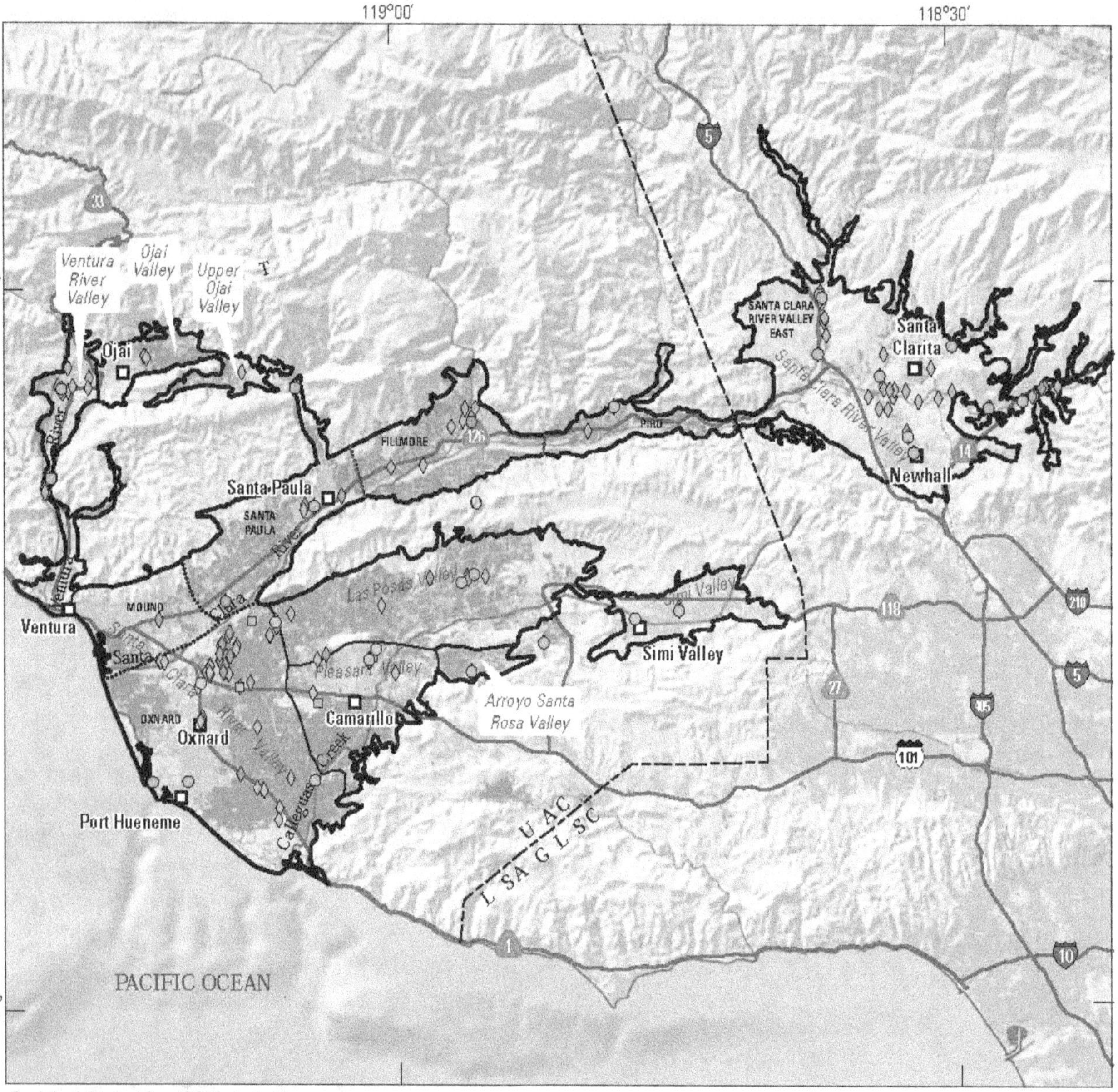

S aded relief derived from U S Geologi al Survey
ational levation Dataset 00
Albers ual Area Coni Pro e tion
ort Ameri an Datum of (AD)

0 5 10 MILES
0 5 10 KILOMETERS

Land-use data from a aga i and ot ers 00

. Gross al a radioa tivity

EXPLANATION

LAN E
LA I I ATION

Urban

Agri ultural

atural

MOUND

(from California De artment of ater esour es 00)

	USGS- or CDPH-grid wells	USGS-understanding wells	California De artment of Publi Healt (CDPH-ot er) wells
Low or not dete ted			
oderate			
Hig			

Figure 12.—Continued

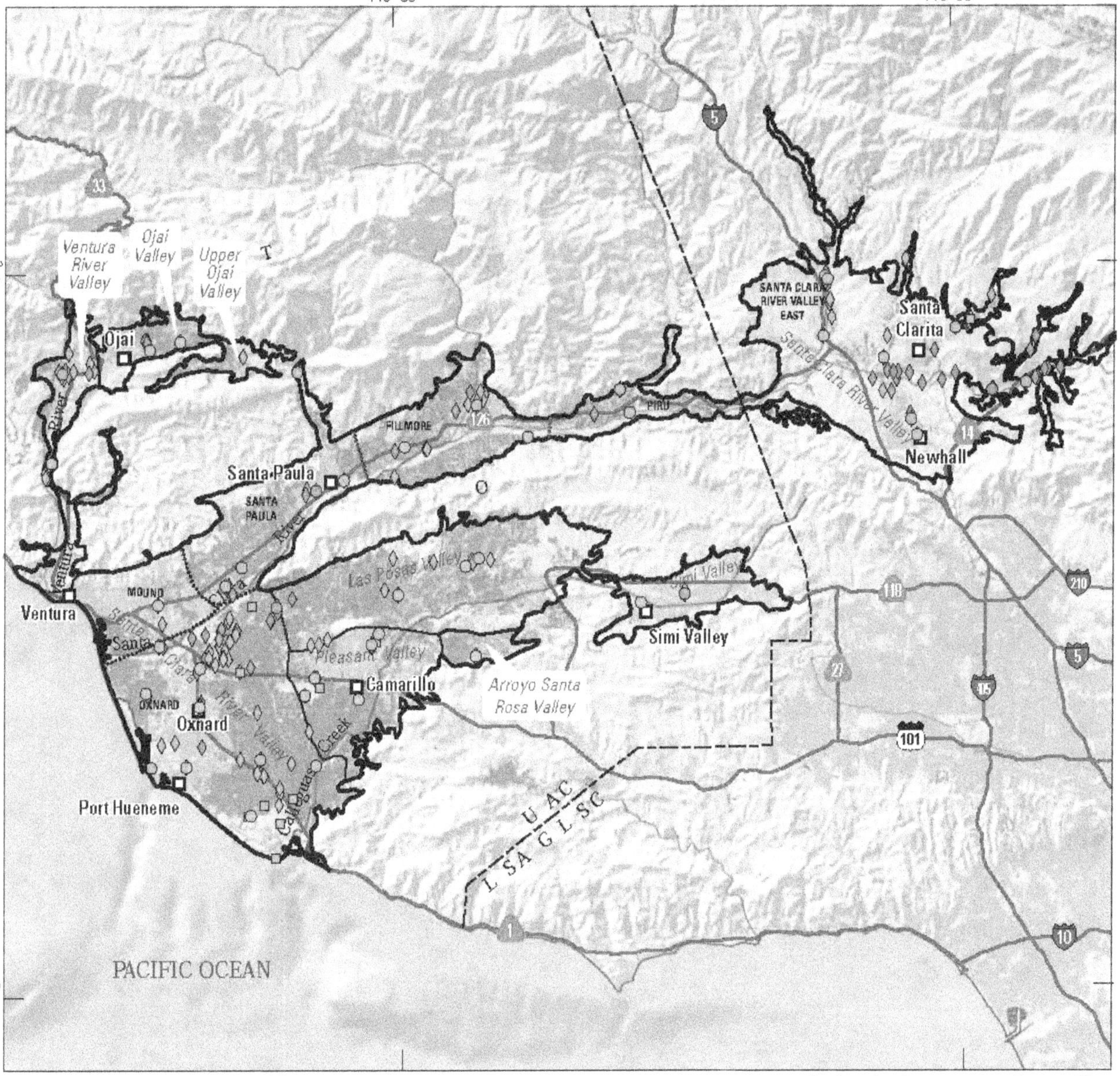

S aded relief derived from U S Geologi al Survey ational levation Dataset 00
Albers ual Area Coni Pro e tion
ort Ameri an Datum of (AD)

0 5 10 MILES
0 5 10 KILOMETERS

Land-use data from a aga i and ot ers 00

. itrate

EXPLANATION

LAN E
LA I I ATION

Urban

Agri ultural

atural

(from California De artment of ater esour es 00)

MOUND

	USGS- or CDPH-grid wells	USGS-understanding wells	California De artment of Publi Healt (CDPH-ot er) wells
Low or not dete ted	○	□	◊
oderate	○	□	◊
Hig	○	□	◊

Figure 12.—Continued

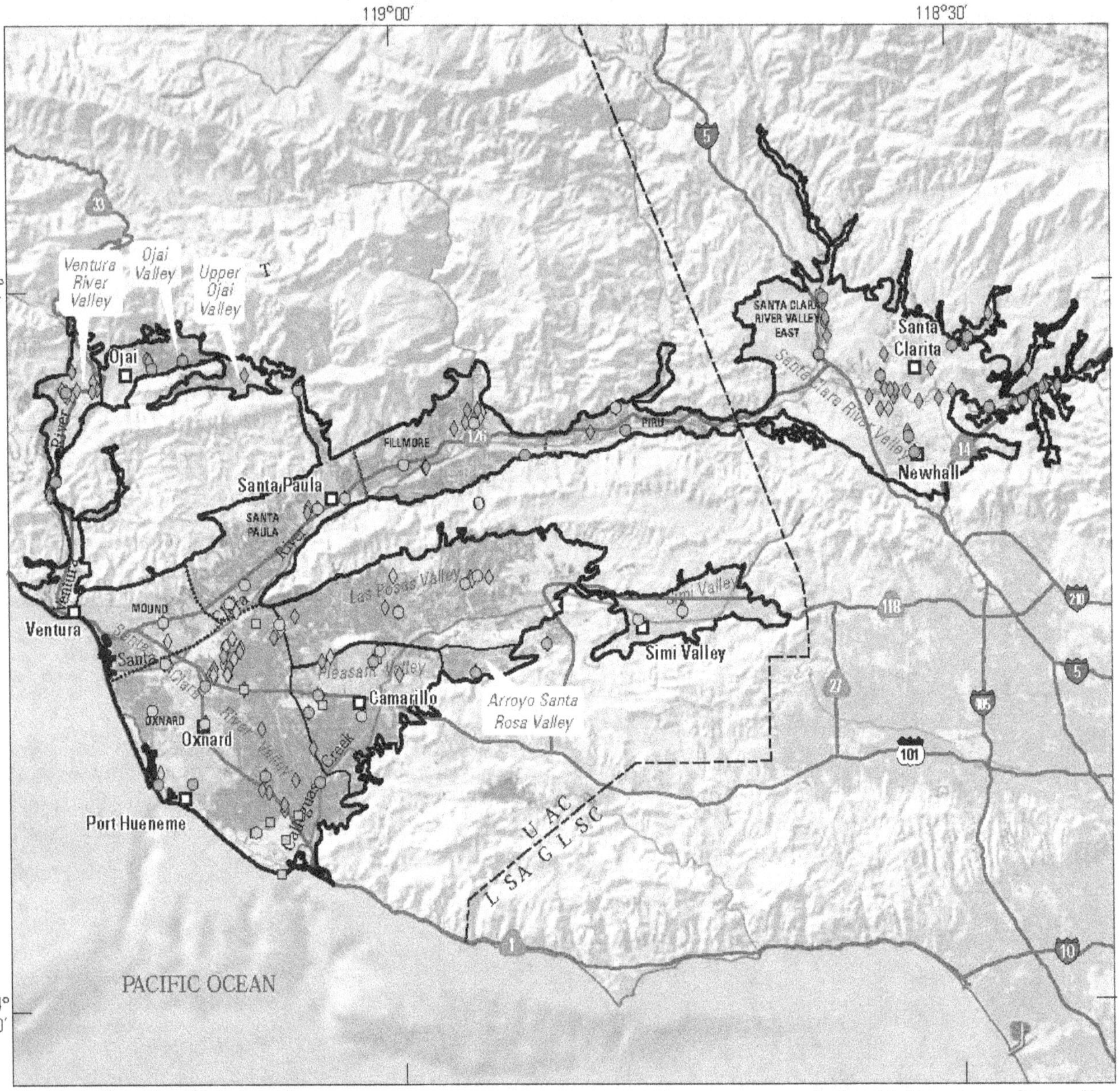

S aded relief derived from U S Geologi al Survey
ational levation Dataset 00
Albers ual Area Coni Pro e tion
ort Ameri an Datum of (AD)

0 5 10 MILES
0 5 10 KILOMETERS

Land-use data from a aga i and ot ers 00

. otal dissolved solids

EXPLANATION

LAN E
LA I I ATION

Urban

Agri ultural

atural

(from California
De artment of ater esour es 00)

MOUND

	USGS- or CDPH-grid wells	USGS-understanding wells	California De artment of Publi Healt (CDPH-ot er) wells
Low or not dete ted	○	□	◊
oderate	○	□	◊
Hig	○	□	◊

Figure 12.—Continued

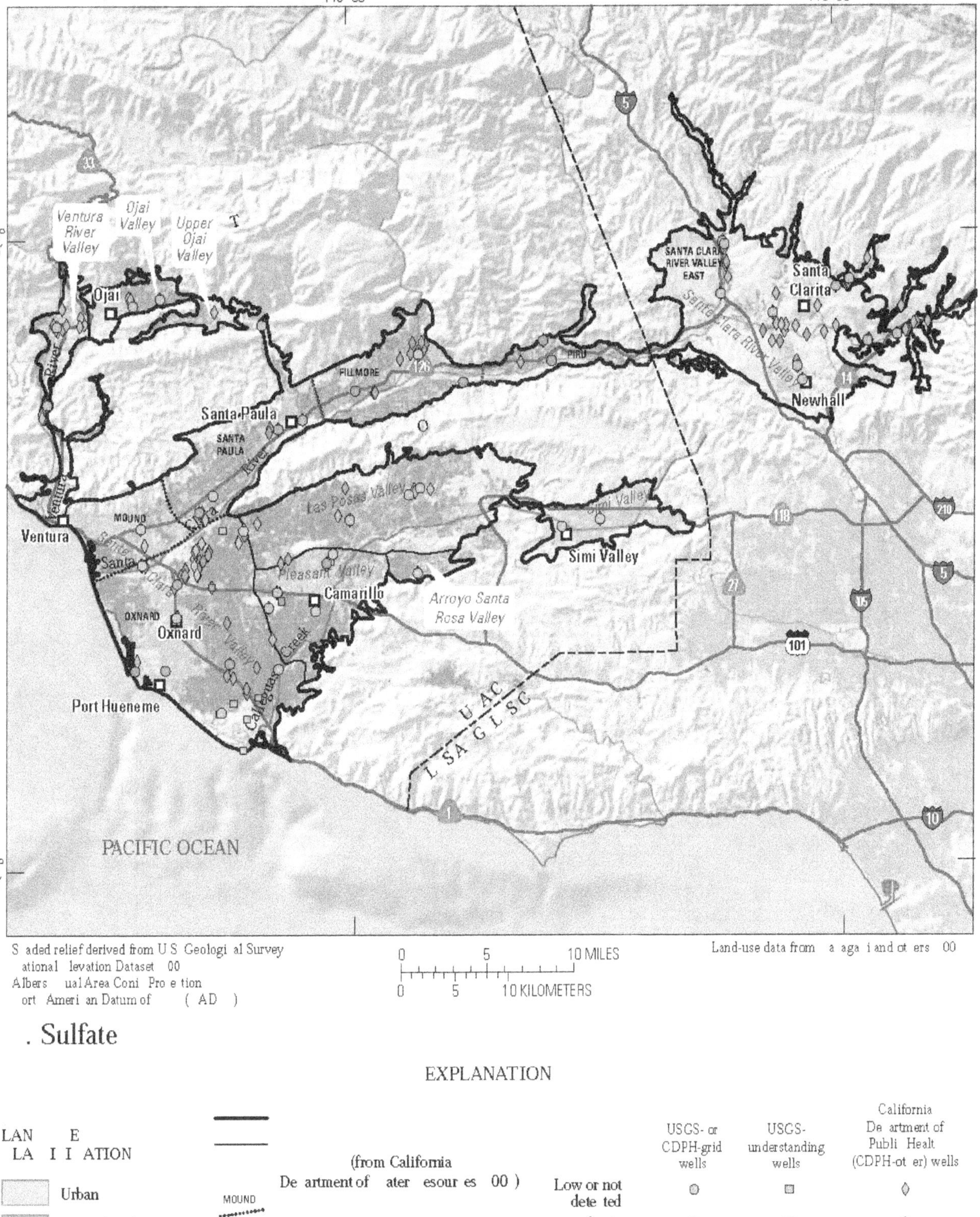

. Sulfate

EXPLANATION

Figure 12.—Continued

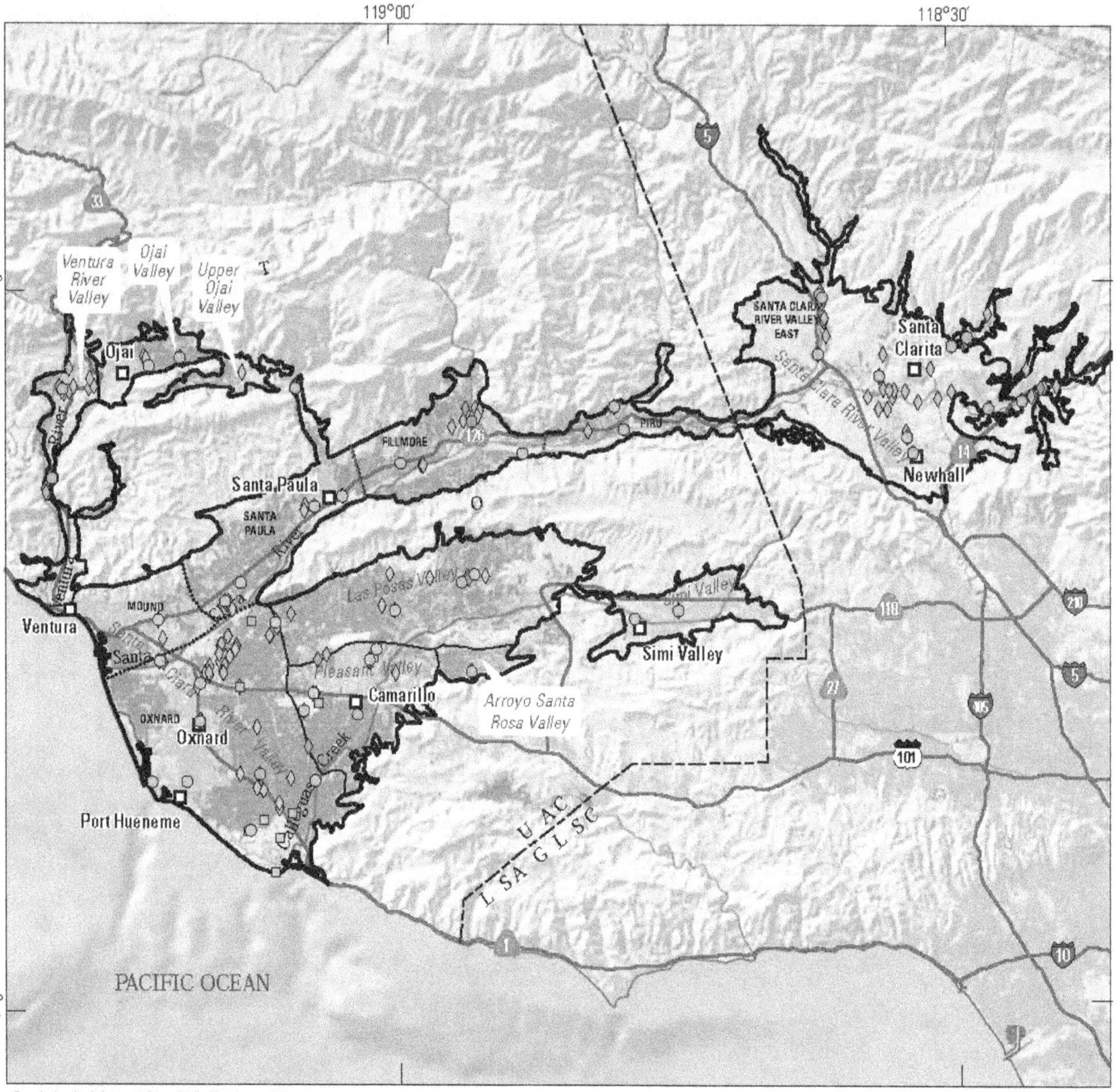

S aded relief derived from U S Geologi al Survey ational levation Dataset 00
Albers ual Area Coni Pro e tion
ort Ameri an Datum of (AD)

0 5 10 MILES
0 5 10 KILOMETERS

Land-use data from a aga i and ot ers 00

. anganese

EXPLANATION

LAN E
LA I I ATION

Urban

Agri ultural

atural

(from California De artment of ater esour es 00)

MOUND

	USGS- or CDPH-grid wells	USGS-understanding wells	California De artment of Publi Healt (CDPH-ot er) wells
Low or not dete ted			
oderate			
Hig			

Figure 12.—Continued

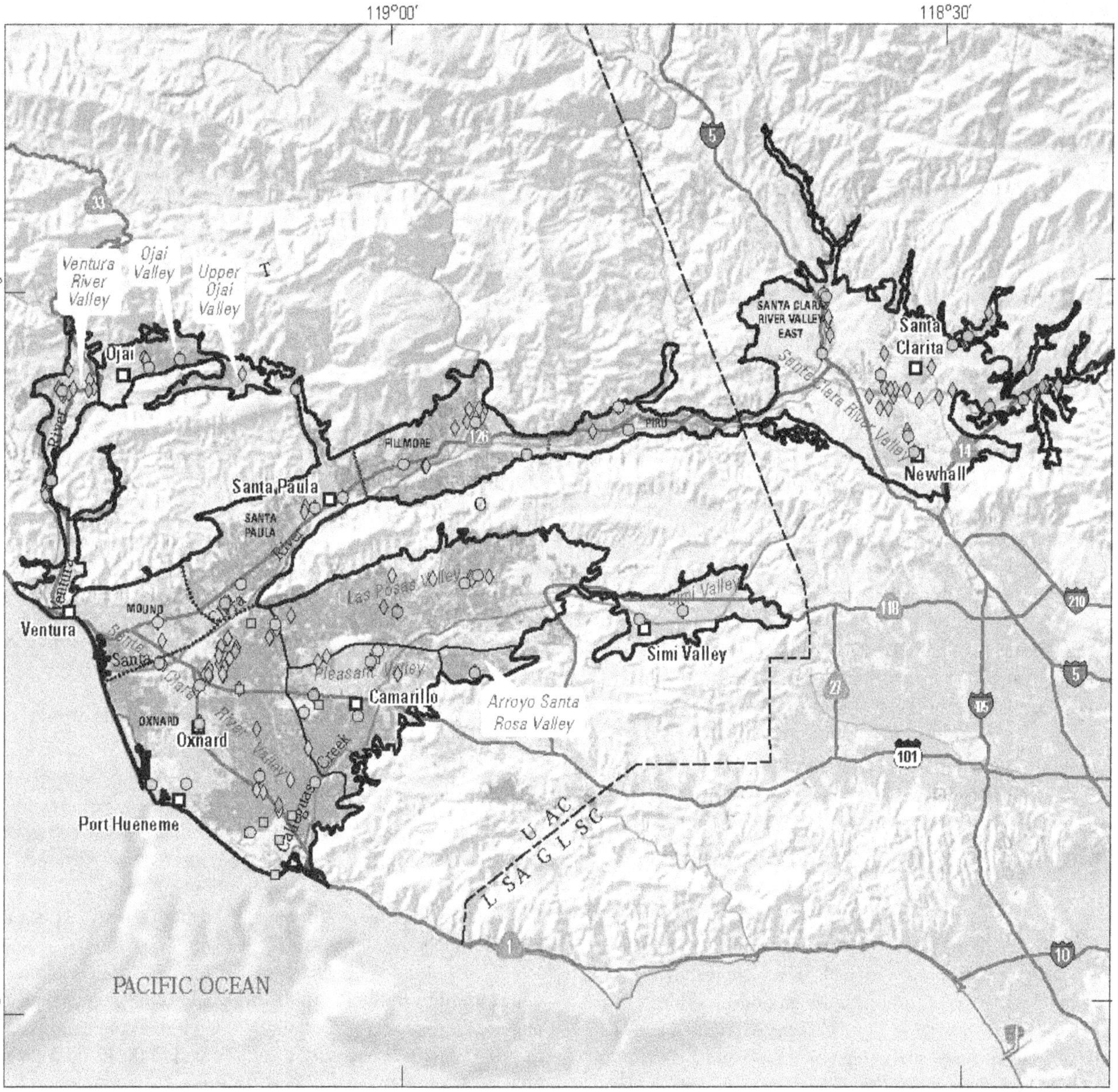

S aded relief derived from U S Geologi al Survey ational levation Dataset 00
Albers ual Area Coni Pro e tion
ort Ameri an Datum of (AD)

0 5 10 MILES
0 5 10 KILOMETERS

Land-use data from a aga i and ot ers 00

. ron

EXPLANATION

LAN E LA I I ATION

Urban

Agri ultural

atural

(from California De artment of ater esour es 00)

MOUND

	USGS- or CDPH-grid wells	USGS-understanding wells	California De artment of Publi Healt (CDPH-ot er) wells
Low or not dete ted	○	□	◊
oderate	○	□	◊
Hig	○	□	◊

Figure 12.—Continued

Factors Affecting Arsenic

Arsenic concentrations in the SCRV were not significantly different between groundwater-age classes (table 6) and were not significantly correlated to any other explanatory factors (table 11). The low frequency of high and moderate relative-concentrations of arsenic in the SCRV may make identification of explanatory factors for arsenic concentrations difficult. An arsenic concentration greater than the MCL-US was detected in one CDPH-grid well (SCRV-DPH-48) in the Santa Clara River Valley East subbasin (fig. 12A). DO data were not available for this CDPH-grid well; the sample pH was 7.8.

Factors Affecting Vanadium

Vanadium concentrations in the SCRV were significantly different between modern and pre-modern groundwater-age classes (table 6) but were not significantly correlated to any explanatory factor (table 11). The low frequency of high and moderate relative-concentrations of vanadium in the Santa Clara River Valley study unit may make identification of explanatory factors for vanadium concentrations difficult. High relative-concentrations of vanadium were detected in one CDPH-grid well (SCRV-DG-13) and two CDPH-other wells located in close proximity to the CDPH-grid well and to each other in the Arroyo Santa Rosa Valley basin (fig. 12B). Desorption of vanadium from aquifer material can occur under oxic conditions (DO >0.5 mg/L) and relatively high pH (>8.0) (Reimann and Caritat, 1998; Wright and Belitz, 2010). The DO and pH of CDPH-grid well SCRV-DG-13 were 2.7 and 7.4 mg/L, respectively, so other processes may account for the high concentration of vanadium in this well.

Factors Affecting Boron

Boron concentrations in the SCRV were negatively correlated with agricultural land use and positively correlated with urban land use (table 11). These correlations indicate that wells surrounded predominantly by urban land use tend to have higher concentrations of boron than wells surrounded by agricultural land use. Boron concentrations were not significantly different between groundwater-age classes (table 6) and were not significantly correlated to any explanatory factor other than land use (table 11).

High boron concentrations in groundwater have been attributed to several sources. Boron can be concentrated in municipal sewage because boron is found in household detergents (washing powders). Because standard treatment processes do not remove boron, it can become concentrated in the treated wastewater effluent (Dotsika and others, 2005). Several wastewater plants are located in the study unit that discharge treated effluent to surface water and (or) allow treated effluent to percolate to groundwater (United Water Conservation District, 2008). In addition to wastewater sources, boron may be found at high concentrations (greater than 1,000 μg/L) in groundwater where the aquifer material is composed of organic-rich sediments (Goodarzi and Swaine, 1994; Williams and Hervig, 2004). High boron concentrations often occur near the coast. It is likely the high concentrations are caused by seawater intrusion because boron concentrations in seawater typically are greater than 4,000 μg/L. High boron concentrations of greater than 2,000 μg/L in understanding wells SRVU-09 and SRVU-10, which are located near the coast, possibly indicate the effects of the mixing of fresh groundwater with seawater in these areas. Additional work is required to more fully explore sources and controls on boron concentrations in the Santa Clara River Valley study unit.

Boron concentrations are greater than the NL-CA in CDPH-other wells in the Fillmore subbasin and Simi Valley basin (fig. 12C). Boron also has been reported as a constituent of concern in the Upper Ojai Valley basin, and in the Piru and Fillmore subbasins (California Department of Public Health, 2004b, 2004h, 2004m). In addition, previous groundwater and surface-water studies have reported boron concentrations greater than the MCL-CA of 1,000 μg/L in the Santa Clara River Valley basin (Izbicki and others, 1995).

Gross Alpha Radioactivity and other Radioactive Constituents

Relative-concentrations of radioactive constituents, as a class, were high in 14 percent, moderate in 11 percent, and low or not detected in 75 percent of the primary aquifer system (table 10). Gross alpha radioactivity was detected at high relative-concentrations in 14 percent and at moderate concentrations in 6.9 percent of the primary aquifer system (table 9, fig. 11). Relative-concentrations of another radioactive constituent, uranium, were moderate in 20 percent of the primary aquifer system (table 9).

Gross alpha radioactivity was detected at high relative-concentrations in the Pleasant Valley and Simi Valley basins and the Oxnard subbasin (fig. 12D). Moderate concentrations of gross alpha radioactivity were detected in the same areas as high concentrations, and in the Las Posas Valley basin, and the Fillmore and Santa Paula subbasins. Gross alpha radioactivity was not significantly different between groundwater-age classes (table 6) and was not significantly correlated to any explanatory factors (table 11).

Nutrients

The relative-concentrations of nutrients, as a class, were high in 15 percent, moderate in 8.7 percent, and low or not detected in 76 percent of the primary aquifer system (table 10). The only nutrient detected at high or moderate relative-concentrations was nitrate plus nitrite (hereinafter referred to as nitrate) (table 9, fig. 10). High or moderate concentrations of nitrate were detected throughout the study unit (fig. 12E).

Table 11. Summary of Spearman's *rho* correlation analysis between selected water-quality constituents and potential explanatory factors, Santa Clara River Valley study unit, California, GAMA Priority Basin Project.

[Aquifer-scale proportions are from the grid-based method unless otherwise stated; *rho* value is shown when the correlations between selected explanatory factors and water-quality constituents are significant ($p < 0.05$); GAMA, Groundwater Ambient Monitoring and Assessment Program; ns, no significant correlation; THM, trihalomethane; TDS, total dissolved solids; MCL-US, U.S. Environmental Protection Agency maximum contaminant level; NL-CA, CDPH notification level; SMCL-CA, CDPH secondary maximum contaminant level; CDPH, California Department of Public Health; <, less than; >, greater than]

Constituent	Benchmark type	Aquifer-scale proportion with high values (percent)	Grid and understanding wells combined: Depth to top-of-perforations	Grid and understanding wells combined: Well depth	Grid and understanding wells combined: pH	Grid and understanding wells combined: Dissolved oxygen	Grid wells: Percentage of land use[1]: Agricultural	Grid wells: Percentage of land use[1]: Urban	Grid wells: Percentage of land use[1]: Natural	Grid wells: Number of septic tanks or cesspools[1]
			rho	*rho*	*rho*	*rho*	*rho*	*rho*	*rho*	*rho*
Inorganic constituents										
Vanadium[2]	NL-CA	3.4	ns	ns	ns	ns	ns	ns	ns	ns
Arsenic[2]	MCL-US	2.3	ns	ns	ns	ns	ns	ns	ns	ns
Boron[2]	NL-CA	[3]3.2	ns	ns	ns	ns	−0.404	2.104	ns	ns
Nitrate[2]	MCL-US	15	−0.618	−0.474	ns	0.575	ns	ns	ns	ns
TDS[2]	SMCL-CA	35	ns	ns	−0.430	ns	ns	ns	ns	ns
Sulfate[2]	SMCL-CA	22	ns	ns	−0.473	ns	ns	ns	−0.341	ns
Manganese	SMCL-CA	38	0.549	0.465	ns	−0.824	ns	ns	ns	−0.376
Iron[2]	SMCL-CA	22	0.436	0.325	ns	−0.391	ns	ns	ns	ns
Radioactive constituents										
Gross alpha radioactivity[2]	MCL-US	14	ns	ns	ns	ns	ns	ns	ns	ns
Organic constituents and constituent classes										
Total THMs[4]	MCL-US	0.0	ns	ns	ns	ns	ns	ns	ns	ns
Solvents, sum of concentrations[4]	variable	0.0	ns	ns	ns	ns	ns	ns	ns	ns
Pesticides, sum of concentrations[4]	MCL-US	0.0	−0.551	−0.481	ns	0.385	ns	ns	ns	ns
Constituents of special interest										
Perchlorate[4]	MCL-US	0.0	ns	ns	−0.292	ns	ns	ns	ns	ns

[1]Land-use percentages and number of septic tanks are within a circle with a radius of 500 meters centered around each well included in analysis.

[2]Constituents with greater than 2 percent high aquifer proportion.

[3]Based on the spatially weighted approach.

[4]Classes of compounds that include constituents with high values less than 2 percent, moderate values, or detection frequencies at any concentration greater than 10 percent.

Factors Affecting Nitrate

Nitrate was negatively correlated with depth to the top-of-perforations and well depth, and positively correlated with DO (table 11, fig. 13). As depth to top-of-perforations, or well depth, increases, the nitrate concentrations decrease (fig. 13*A*). Nitrate concentrations also were significantly greater in wells with modern or mixed groundwater ages than in wells with pre-modern groundwater ages (table 6, fig. 13*A*). The occurrence of high nitrate and DO concentrations in shallow and younger groundwater indicates surficial or near-surface sources of nitrate. The distribution of nitrate concentrations in SCRV was expected and is consistent with the distribution of nitrate concentrations observed in other aquifers (Hallberg and Keeney, 1993; Burow and others, 2007; Landon and others, 2010). Some of the explanatory variables related to nitrate concentrations are themselves related; for example, DO and groundwater age are both correlated to depth to the top-of-perforations (tables 6 and 7, fig. 9*A*).

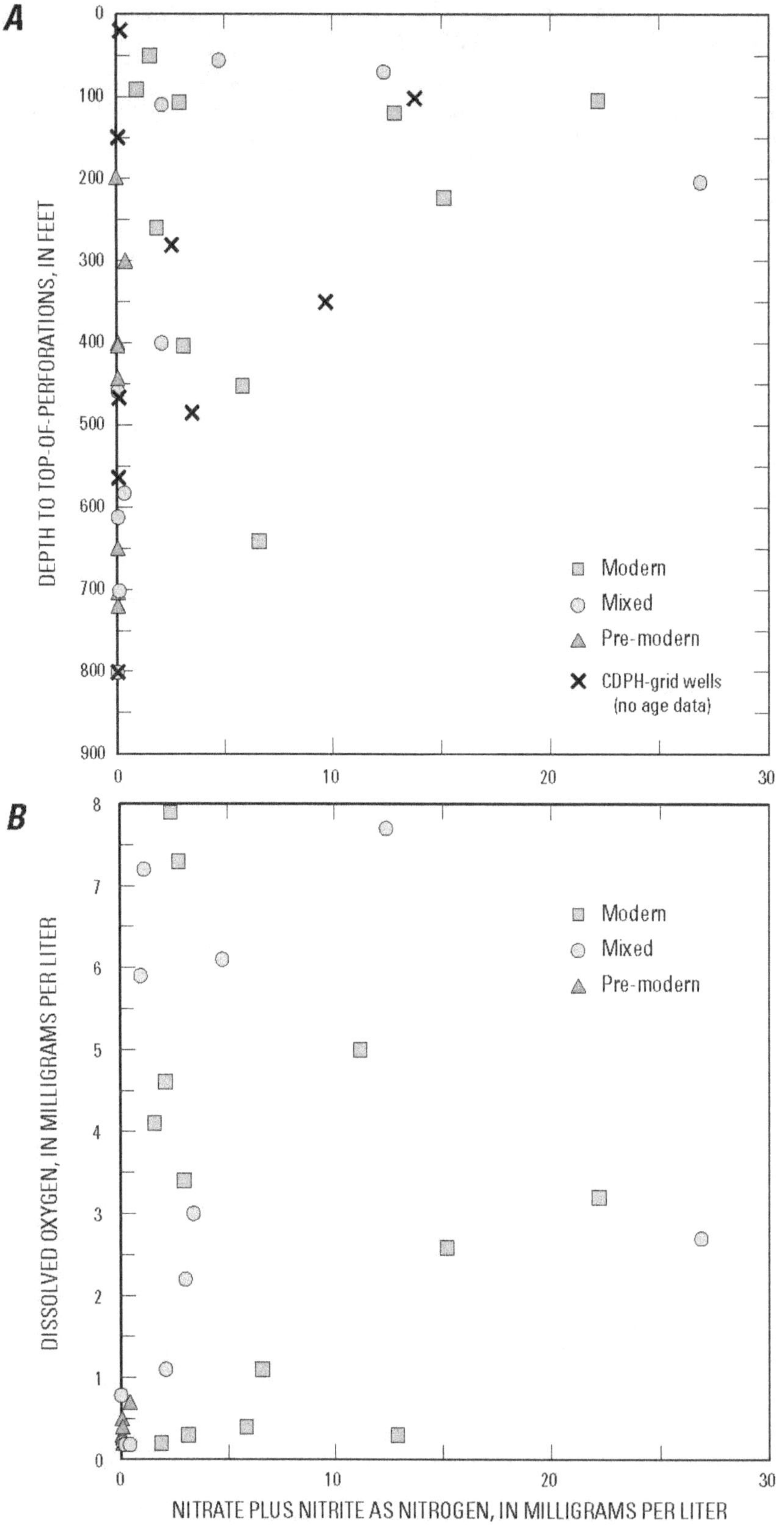

Figure 13. Relation of nitrate concentrations to depth to top-of-perforations and dissolved-oxygen concentrations classified by groundwater age for grid and understanding wells, Santa Clara River Valley study unit, California GAMA Priority Basin Project.

Available information indicates that denitrification may contribute to decreasing nitrate concentrations at increasing depths, where anoxic conditions occur. As DO concentrations in groundwater decrease with depth (fig. 9A), nitrate becomes the next most favorable electron acceptor; therefore, microbially mediated denitrification of nitrate to dissolved nitrogen gas (N^2) and to intermediate products can occur (Kendall, 1998). Previous groundwater studies in the Santa Clara River Valley study unit have confirmed that denitrification occurs in parts of the aquifer (Izbicki and others, 2005). Under the generally widespread anoxic conditions in the deep aquifers in the Santa Clara River Valley study unit, the distribution of nitrate in the aquifer generally is expected to be influenced by denitrification.

Relations between nitrate concentrations and agricultural land use or the density of septic tanks or cesspools surrounding grid wells did not meet the criteria for being significant (table 11). The land use within a 500-m radius surrounding the seven grid wells that had a nitrate concentration greater than the MCL-US of 10 mg/L varied from predominantly agricultural or urban land use to mixed land use (fig. 12*E*). Relations between agricultural land use and nitrate concentrations (for example, Kendall, 1998; Burow and others, 2007) and between septic systems and nitrate concentrations (for example, Moore and others, 2006) have been noted in other aquifers. The absence of relations between nitrate concentration and land use in this study unit may reflect multiple factors, including many sources of nitrate, generally mixed land uses on the landscape, or mixing of water and solutes from multiple land uses, especially in relatively deep wells with long perforation intervals.

Inorganics with Aesthetic Benchmarks (SMCL)

As a class, inorganics with aesthetic benchmarks (SMCLs) had high relative-concentrations in 54 percent, moderate in 41 percent, and low in 4.3 percent of the primary aquifer system (table 10). The constituents detected at high relative-concentrations were TDS, sulfate, manganese, and iron (table 9, figs. 10 and 11). Chloride was detected at moderate relative-concentrations in 2.2 percent of the primary aquifer system (table 9).

TDS was detected at high relative-concentrations in 35 percent and moderate in 56 percent of the primary aquifer system (table 9). TDS relative-concentrations were high throughout the Santa Clara River Valley basin but were more prevalent in the downgradient areas of the Santa Clara River Valley (closest to the coastline), as well as in the Pleasant Valley and Simi Valley basins (fig. 12*F*).

Sulfate was detected at high relative-concentrations in 22 percent and moderate in 27 percent of the primary aquifer system (table 9). Sulfate relative-concentrations were high and moderate in the downgradient areas of the Santa Clara River Valley basin (closest to the coastline), as well as in the Pleasant Valley and Simi Valley basins (fig.12*G*).

Manganese was detected at high relative-concentrations in 38 percent and moderate in 2.2 percent of the primary aquifer system (table 9). Manganese relative-concentrations generally were high in the downgradient areas of the Santa Clara River Valley, and in the Las Posas Valley, Pleasant Valley, Ojai Valley, and Ventura River Valley basins (fig. 12*H*).

Iron was detected at high relative-concentrations in 22 percent and moderate in 4.4 percent of the primary aquifer system (table 9). The distribution of high relative-concentrations of iron was similar to manganese, with high concentrations in the downgradient areas of the Santa Clara River Valley and in the Las Posas Valley and Pleasant Valley basins (fig. 12*I*). However, at least one relative-concentration also was high in the Piru and Santa Clara River Valley East subbasins and in the Ventura River Valley basin (fig. 12*I*).

Factors Affecting Total Dissolved Solids (TDS)

TDS concentrations were negatively correlated with pH (table 11), indicating that TDS concentrations generally decrease as pH increases (fig. 14*A*). In general, groundwater in wells with high TDS (>800 mg/L) and low pH (<7.5) was modern or mixed ages, whereas groundwater in wells with low TDS (≤800 mg/L) and high pH (≥7.5) more commonly was pre-modern age (fig. 14*A*). There are exceptions to these general relations of TDS, pH, and age. TDS concentrations for four understanding wells (SCRVU-07, -08, -09, -10) in the southwestern corner of the study unit (fig. A1*A*) near the coast were 8,740 to 30,000 mg/L. The chemistry of these wells indicates the effects of mixing with seawater or the upwelling of brines from deep formations (Izbicki and others, 2005).

In general, TDS concentrations are higher in wells in the downgradient parts of the regional groundwater flow system of the study unit (fig. 12*F*), for example, in the downgradient areas of the Santa Clara River Valley and Pleasant Valley basins. TDS concentrations greater than the SMCL-CA of 1,000 mg/L were detected in only two grid wells in Simi Valley basin, which could be considered in the upgradient area of the study unit. Marine sediments surround and underlie Simi Valley (fig. 5) and could potentially affect the TDS concentration in the groundwater where these wells are located. The proportions of the primary aquifer system with high and moderate relative-concentrations are larger in SCRV than in many of the study units evaluated by the Priority Basin Project (Bennett and others, 2010; Kulongoski and others, 2010; Landon and others, 2010; Bennett and others, 2011).

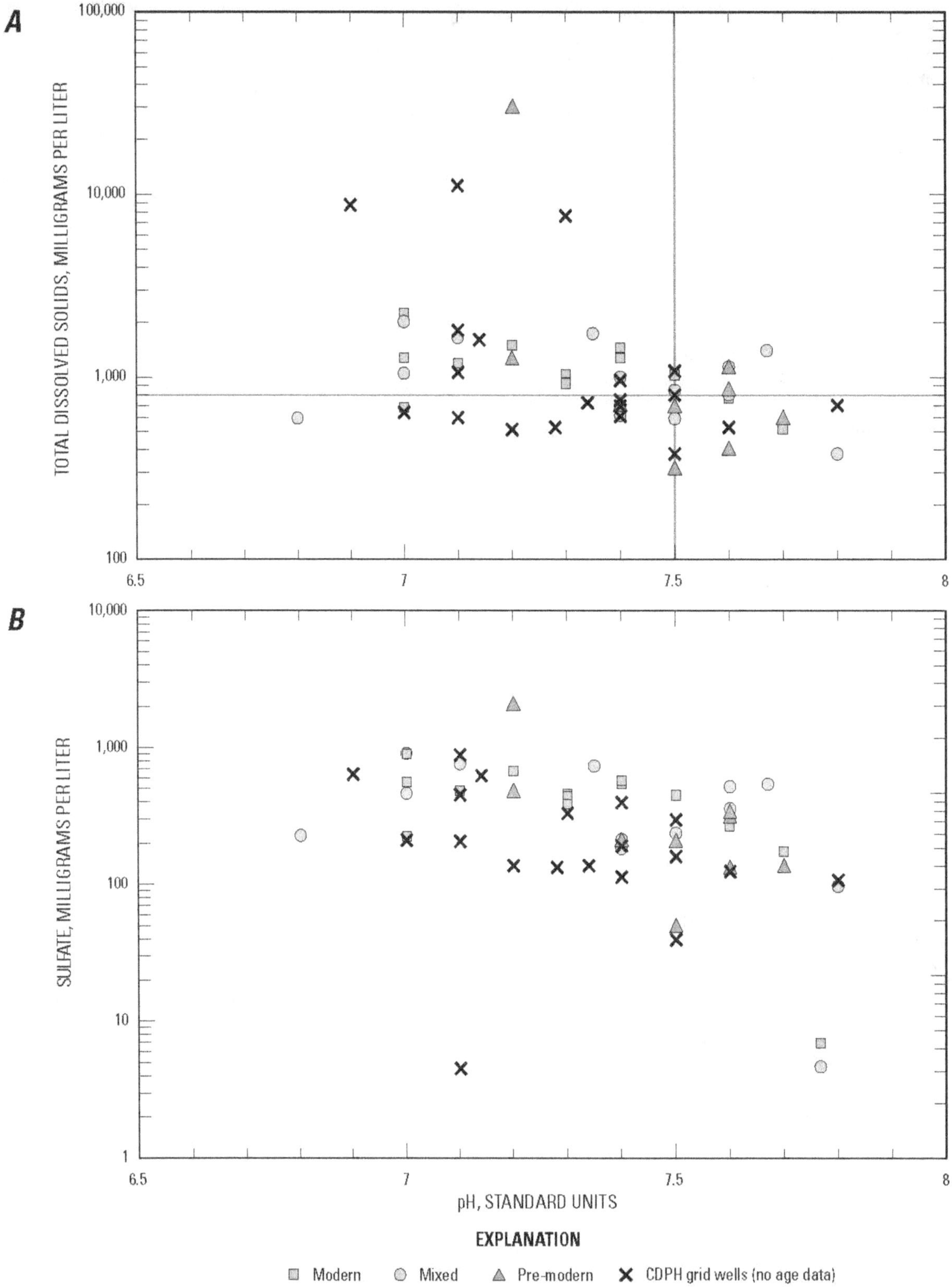

Figure 14. Relations of total dissolved solids and sulfate concentrations to pH, classified by groundwater age, for grid and understanding wells, Santa Clara River Valley study unit, California GAMA Priority Basin Project.

Concentrations of TDS and sulfate were strongly positively correlated (ρ = 0.775, not shown in table 10), and high relative-concentrations of these two constituents often co-occurred (figs. 12*F* and 12*G*). The correlation of TDS and sulfate was expected because sulfate is one of the two most abundant anions in groundwater in the study unit (appendix B, fig. B2). Moreover, sulfate typically was the most abundant anion in groundwater with high TDS concentrations because the concentration of the other major anion, bicarbonate, is limited by precipitation of calcite at high concentrations (Izbicki and others, 2005).

The high and moderate TDS concentrations in the SCRV study unit could indicate a variety of factors, both natural and anthropogenic, including historical groundwater pumping and evapotranspiration patterns, irrigation return and irrigation recycling, mixing of shallow, fresh groundwater with deep, more saline groundwater influenced by interactions with deep marine sediments, intrusion of seawater or upwelling of deep saline waters, and rock/water interaction along regional groundwater flow paths. High salinity water is present in the perched and semi-perched aquifers and also in the deep aquifer zones in parts of the study unit (Izbicki, 1996a; Izbicki and others, 2005). Further discussion of sources of salinity in the SCRV study unit is beyond the scope of this report.

Factors Affecting Sulfate

Similar to TDS, sulfate concentrations were negatively correlated with pH (table 11), indicating that sulfate concentrations decrease as pH increases (fig. 14*B*). Sulfate concentrations were not significantly different between groundwater age classes (table 6).

Similar to TDS, the source of sulfate can vary across the study unit. For example, a core sample from relatively shallow clay layers (about 130 ft below land surface) underlying the Santa Paula subbasin contained pore water with sulfate concentrations of 3,000 mg/L (Reichard and others, 1999). High concentrations of sulfur may be attributed to the presence of Miocene and Pliocene marine sediments from the surrounding mountains (fig. 5). Izbicki and others (2005) analyzed the sulfuric isotopic composition (δ34S) of sulfate (SO_4) from different depths of the aquifer in the Pleasant Valley basin and in the southwestern part of the Santa Clara River Valley basin to evaluate the source of sulfate in groundwater. Izbicki and others (2005) concluded that the sources of elevated sulfate include (1) evaporative irrigation return water, (2) oxidation of reduced sulfur in sulfide minerals (for example, pyrite) and (or) organic matter in sediments, and (3) dissolution of marine evaporite minerals (for example, gypsum). The concentrations of sulfate in different parts of the study unit vary depending on the depth at which samples were collected and the redox condition of the groundwater. Sulfate-reducing conditions, which result in consumption of sulfate and generation of hydrogen sulfide gas, have been documented to occur widely, particularly in deep parts of the aquifer (Izbicki and Martin, 1997; Izbicki and others, 2005). Precipitation of metallic sulfide minerals, such as pyrite also may occur in parts of the aquifer (Izbicki and others, 2005). The sources and sinks for sulfate determined in these previous investigations are consistent with the decrease in sulfate concentrations with increasing pH.

The negative correlation of sulfate with percentage of natural land use (table 11) may not indicate a direct relation to land use but rather that geochemical processes are producing high sulfate concentrations in the downgradient end of the study unit, where land use is dominantly agricultural or urban. The moderate and low relative-concentrations of sulfate along the coastal areas (near Port Hueneme and the southern tip of the Oxnard subbasin) of the study unit (fig. 12*G*) may indicate sulfate-reduction processes occurring in the aquifer, with the exception of well SCRVU-10. SCRVU-10 is a shallow monitoring well on the coast near the southern tip of the Oxnard subbasin that is affected by seawater intrusion (Izbicki and others, 2005). Public-supply wells in this area have been inactive or on standby because of high TDS or major-ion concentrations as a result of seawater intrusion or upwelling of brine from underlying sediments (Izbicki and others, 2005).

Factors Affecting Manganese

Manganese concentrations primarily vary in response to redox conditions and depth in the SCRV study unit. Manganese concentrations were negatively correlated with DO (table 11, fig. 15*B*). DO concentrations less than 0.5 mg/L and manganese concentrations greater than 50 μg/L generally are consistent with reducing aquifer conditions (predominant redox process is manganese-reduction, table D4). Manganese concentrations were positively correlated with well depth and the depth to top-of-perforations (table 11), indicating increasing concentrations with increasing depth (fig. 15*A*, y-axis scale is reversed so depths increase from top to bottom of plot). These correlations are consistent with increasingly reduced conditions (decreasing DO) in the groundwater system with increasing depth. Manganese concentrations were not significantly different between groundwater age classes (table 6). Most of the wells with manganese concentrations greater than the SMCL-CA of 50 μg/L (or high relative-concentrations) are in the downgradient area of the Santa Clara River Valley basin, and in the Pleasant Valley, Las Posas Valley, Ventura River Valley, or Ojai basins (fig. 12*H*). The negative correlation of manganese concentrations with number of septic systems or cesspools (table 11) likely is not a direct relation but may be an artifact of relations between explanatory factors.

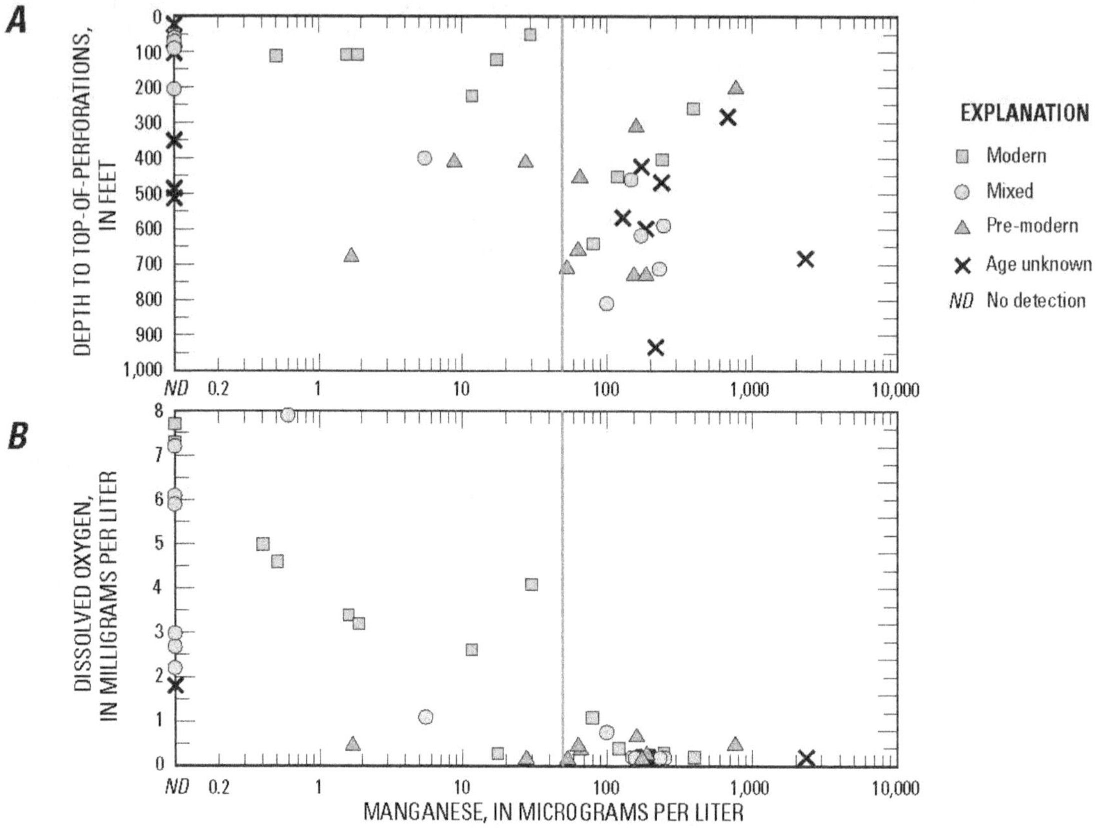

Figure 15. Relation of manganese concentrations to depth to the top-of-perforations and to dissolved-oxygen concentrations, classified by groundwater age, in grid and understanding wells in the Santa Clara River Valley study unit, California GAMA Priority Basin Project.

Factors Affecting Iron

Similar to manganese, iron concentrations were negatively correlated with DO (table 11, fig.16D), and positively correlated well depth (table 11) and the depth to top-of-perforations (table 11, fig.16C), indicating increasing concentrations with increasing depth. Concentrations were higher in groundwater samples with pre-modern than in groundwater samples with modern ages (table 6, figs.16C and 16D). Most of the wells with iron concentrations greater than the SMCL-CA of 300 µg/L (or high relative-concentrations) are in the Oxnard, Mound, Fillmore, and Santa Paula subbasins of the Santa Clara River Valley basin, or in the Pleasant Valley and Simi Valley basins (fig. 12I). For most samples with iron concentrations greater than the SMCL-CA of 300 µg/L (high relative-concentration), manganese concentrations also were greater than the SMCL-CA of 50 µg/L

Iron concentrations may decrease in deep parts of the primary aquifers under anoxic conditions. Sulfate reducing and methanogenesis conditions are present in the deep parts (typically greater than 800 ft below land surface) of the Lower-aquifer system of the Pleasant Valley basin and Oxnard subbasin (Izbicki and others, 1995, 2005). These anoxic conditions persist in these aquifers because of the abundance of electron donors, usually in the form of dissolved organic carbon (DOC), present in marine sediments that makeup the aquifer material. If redox conditions for sulfate reduction exist, then sulfide is produced, which reacts with the dissolved Mn2+ and Fe2+ ions. This reaction can lead to the precipitation of Mn- and Fe-sulfide minerals (Izbicki and others, 2005), thus contributing to decreasing iron concentrations.

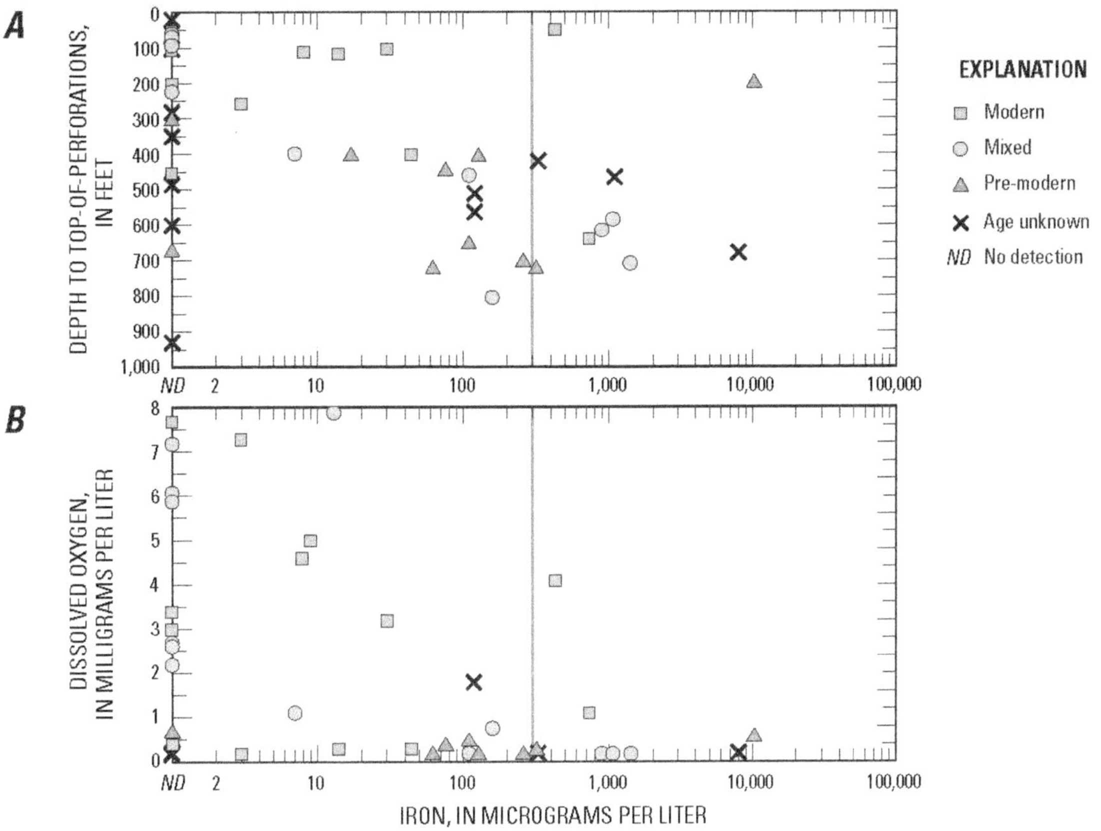

Figure 16. Relation of iron concentrations to depth to top-of-perforations and dissolved-oxygen concentrations, classified by groundwater age, in grid and understanding wells in the Santa Clara River Valley study unit, California GAMA Priority Basin Project.

Organic and Special-Interest Constituents

Organic and special-interest constituents, unlike inorganic constituents, usually are of anthropogenic origin. The organic and special interest constituents discussed in this report include VOCs, pesticides, and perchlorate. VOCs may be present in paints, solvents, fuels, refrigerants, can be byproducts of water disinfection, and are characterized by their tendency to evaporate. In this report, VOCs are classified as trihalomethanes (THMs), solvents, and other VOCs. Pesticides are used to control weeds, insects, or fungi in agricultural, urban, and suburban settings. The only special-interest constituent of concern is perchlorate because perchlorate has been detected recently in, or is considered to have the potential to reach, water resources used for drinking-water supply (California Department of Public Health, 2008b).

Organic constituents with human-health benchmarks, as a whole, were not detected at high relative-concentrations in the primary aquifer system and were detected at moderate relative-concentrations (>0.1 but ≤1) in 4.8 percent of the primary aquifers (table 10). The maximum relative-concentrations for most of the detected organic constituents with human-health benchmarks were less than 0.1 (fig. 17). Organic and special-interest constituents with one or more moderate relative-concentrations were the solvents carbon tetrachloride and trichloroethene (TCE), total THMs, and the special-interest constituent perchlorate (fig. 17). A few organic constituents—THM chloroform, pesticides atrazine and simazine, and the special-interest constituent perchlorate—were prevalent (detection frequency greater than 10 percent in USGS-grid wells) in the primary aquifers (figs. 17 and 18), but at concentrations that were less than their respective benchmarks.

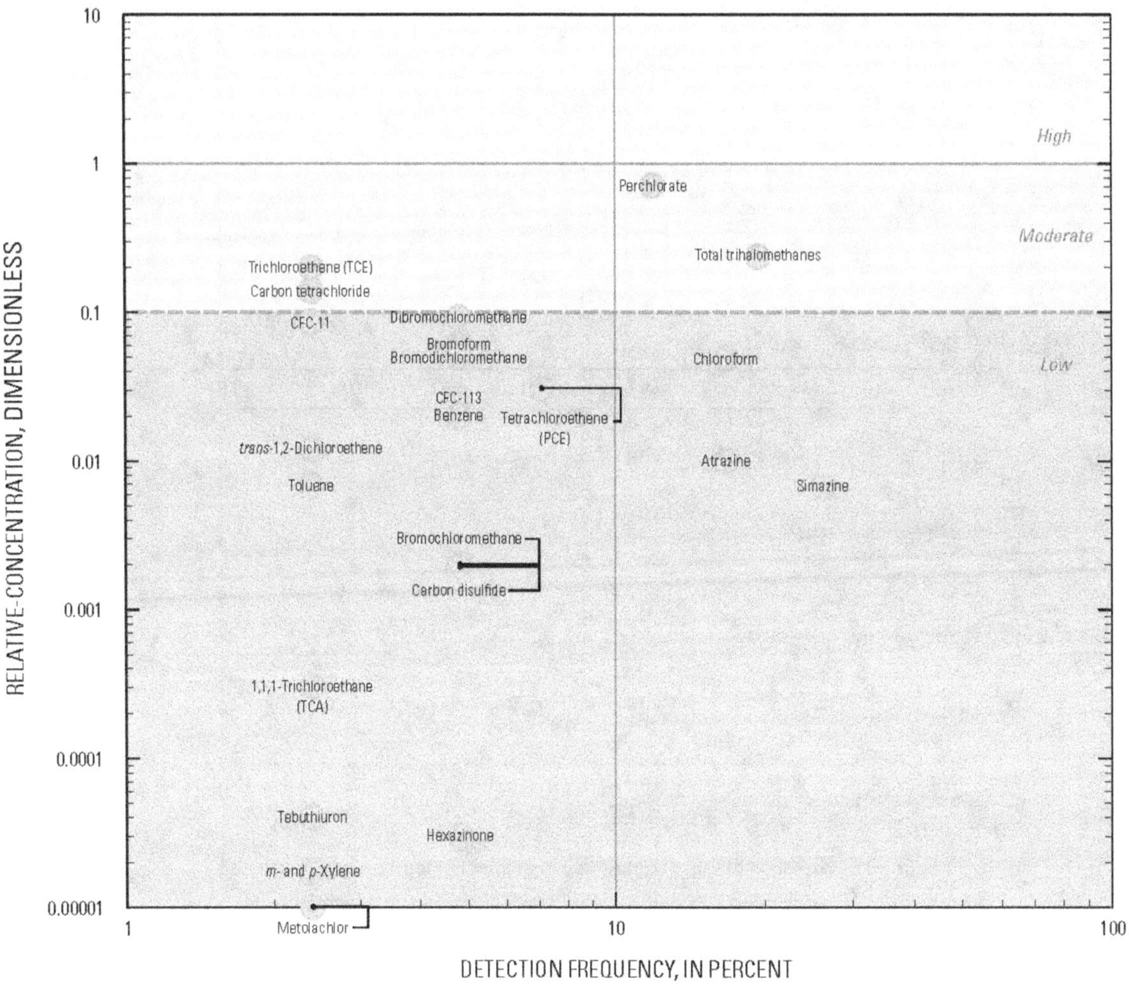

EXPLANATION

Toluene **Constituents with analyses in ≥24 grid wells and wells are spatially representative**

Name and center of symbol is location of data unless indicated by following location line:

Figure 17. Detection frequency and maximum relative-concentration for organic and special-interest constituents detected in USGS-grid wells, Santa Clara River Valley study unit, California GAMA Priority Basin Project.

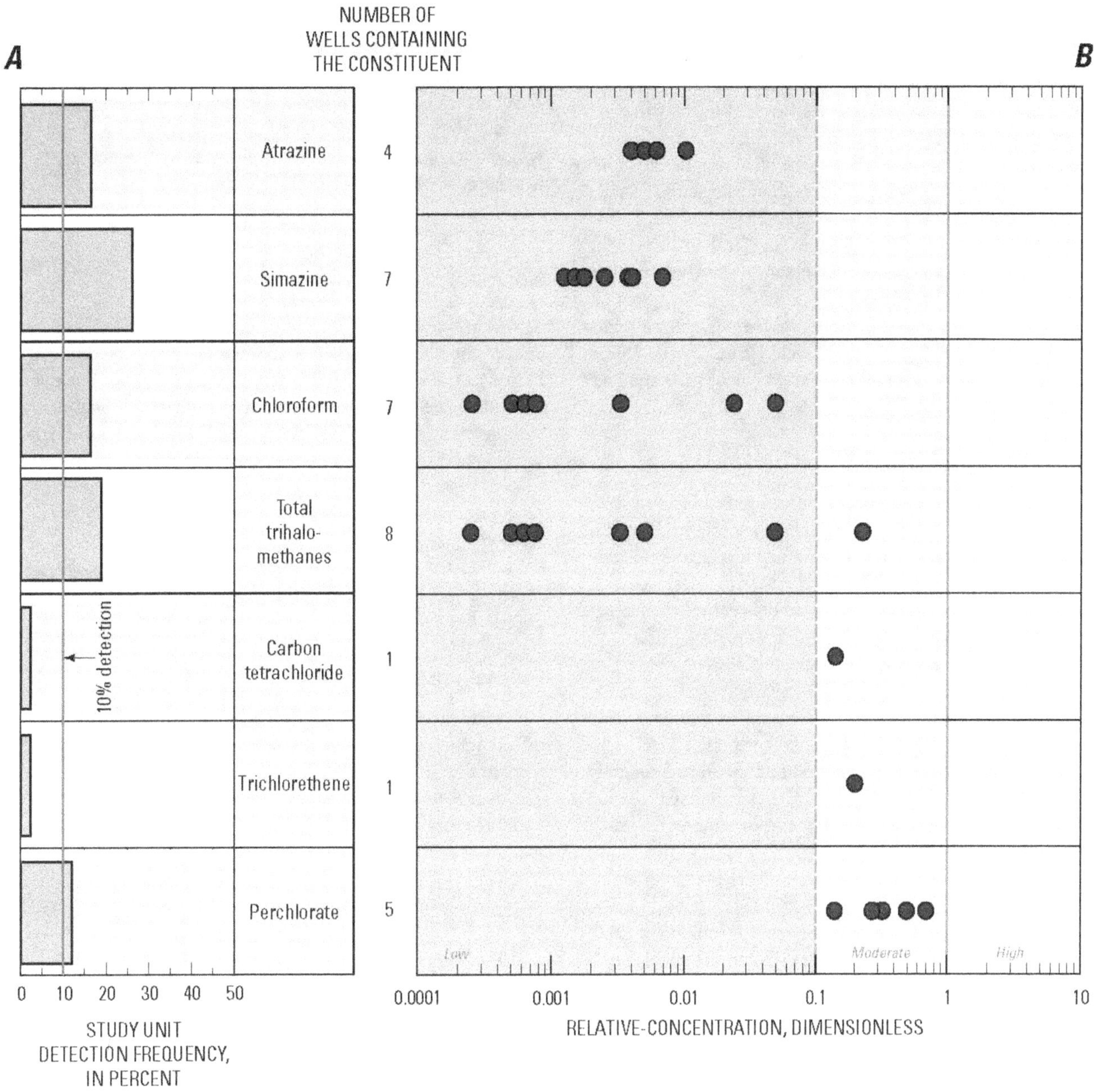

Figure 18. (*A*) Detection frequency and (*B*) relative-concentration of selected organic constituents and perchlorate in USGS-grid wells in the Santa Clara River Valley study unit, California GAMA Priority Basin Project.

Pesticides

Pesticides, as a class, were detected at low relative-concentrations or were not detected in 100 percent of the primary aquifer system (table 10). Most grid wells with detections had one to four pesticides or degradates detected; however, there were two wells that had five or more pesticides detected (fig. 19*A*). Several of the pesticide detections were located in the Santa Clara River Valley East subbasin. The pesticides with detection frequencies greater than 10 percent were the herbicides simazine and atrazine. Simazine was the most frequently detected pesticide in SCRV, with a detection frequency of 26 percent. The detection frequency for atrazine was 17 percent (fig. 18) (Montrella and Beltiz, 2009). The relative-concentration for all concentrations of simazine and atrazine was less than 0.01 (fig. 18).

Factors Affecting Pesticides

As a class, pesticide concentrations were higher in shallower groundwater than in deeper groundwater. Pesticides were negatively correlated with well depth and depth to top-of-perforations and positively correlated with DO (table 11). These relations indicate that pesticide concentrations decrease with increasing depth of well perforations (fig. 20). The correlation of DO and pesticide concentrations probably indicates a parallel decrease in DO with increasing depth (fig. 9*A*). The sum of pesticide concentrations was not significantly different between groundwater age classes (table 6). However, pesticides were detected more frequently in groundwater with modern (45 percent) and mixed (46 percent) ages than in groundwater of pre-modern age (10 percent, fig. 20).

Pesticide concentrations were not correlated with land use in the Santa Clara River Valley study unit (table 11), as has been observed in some studies (Gilliom and others, 2006). Pesticide concentrations were not correlated with land use even in wells with depths to top-of-perforations less than 250 ft. This result was not unexpected because pesticides, in general, are used in both agricultural and urban areas (Gilliom and others, 2006).

Trihalomethanes

THMs, as a class, were detected at moderate relative-concentrations in 2.4 percent and low or not detected in 98 percent of the primary aquifer system (table 10). Chloroform was the most prevalent THM with a detection frequency of 17 percent (fig. 18) and was the most frequently detected VOC in the SCRV (Montrella and Belitz, 2009). Some THMs were detected in most basins and subbasins in the SCRV (fig. 19*B*).

Factors Affecting Trihalomethanes

Total THM concentrations were not correlated with any explanatory variables (table 11). Total THM concentrations were not significantly different between groundwater age classes (table 6). The absence of correlations of total THM with explanatory variables may reflect the low detection frequency for total THM, making it difficult to identify relations.

Solvents

Solvents, as a class, were detected at moderate relative-concentrations in 2.4 percent, and low or not detected in 98 percent of the primary aquifer system (table 10). The two solvents detected at moderate relative-concentrations were carbon tetrachloride and TCE (fig. 17). None of the solvents were detected in more than 10 percent of the grid wells.

Carbon tetrachloride and TCE were detected at moderate relative-concentrations (fig. 18) in one USGS-grid well in the Arroyo Santa Rosa Valley basin (fig. 19*C*). Other solvents were detected at low relative-concentrations in a few USGS-grid wells across the SCRV (fig. 19*C*) (Montrella and Belitz, 2009).

Factors Affecting Solvents

Detections of solvents, as a class, were not significantly different between groundwater age classes (table 6) and also were not correlated with any explanatory variables (table 11). The absence of correlations of solvents with explanatory variables may reflect the low detection frequency for solvents, making it difficult to identify relations.

Other Volatile Organic Compounds (VOCs)

For the other VOCs, relative-concentrations were neither high nor moderate in any proportion of the primary aquifer system based on the grid wells (table 10). The relative-concentrations also were neither high nor moderate in CDPH-other wells during the current period. None of the other VOCs were detected in more than 10 percent of the grid wells.

Special-Interest Constituent

Perchlorate was the only constituent of special interest sampled for in the SCRV. Perchlorate was detected at moderate relative-concentrations in 12 percent of the primary aquifer system (table 9, fig. 17). All perchlorate detections were at moderate relative-concentrations (fig. 17). Perchlorate was detected in the Ojai Valley, Simi Valley, and Arroyo Santa Rosa Valley basins as well as in the Fillmore subbasin (fig. 19*D*).

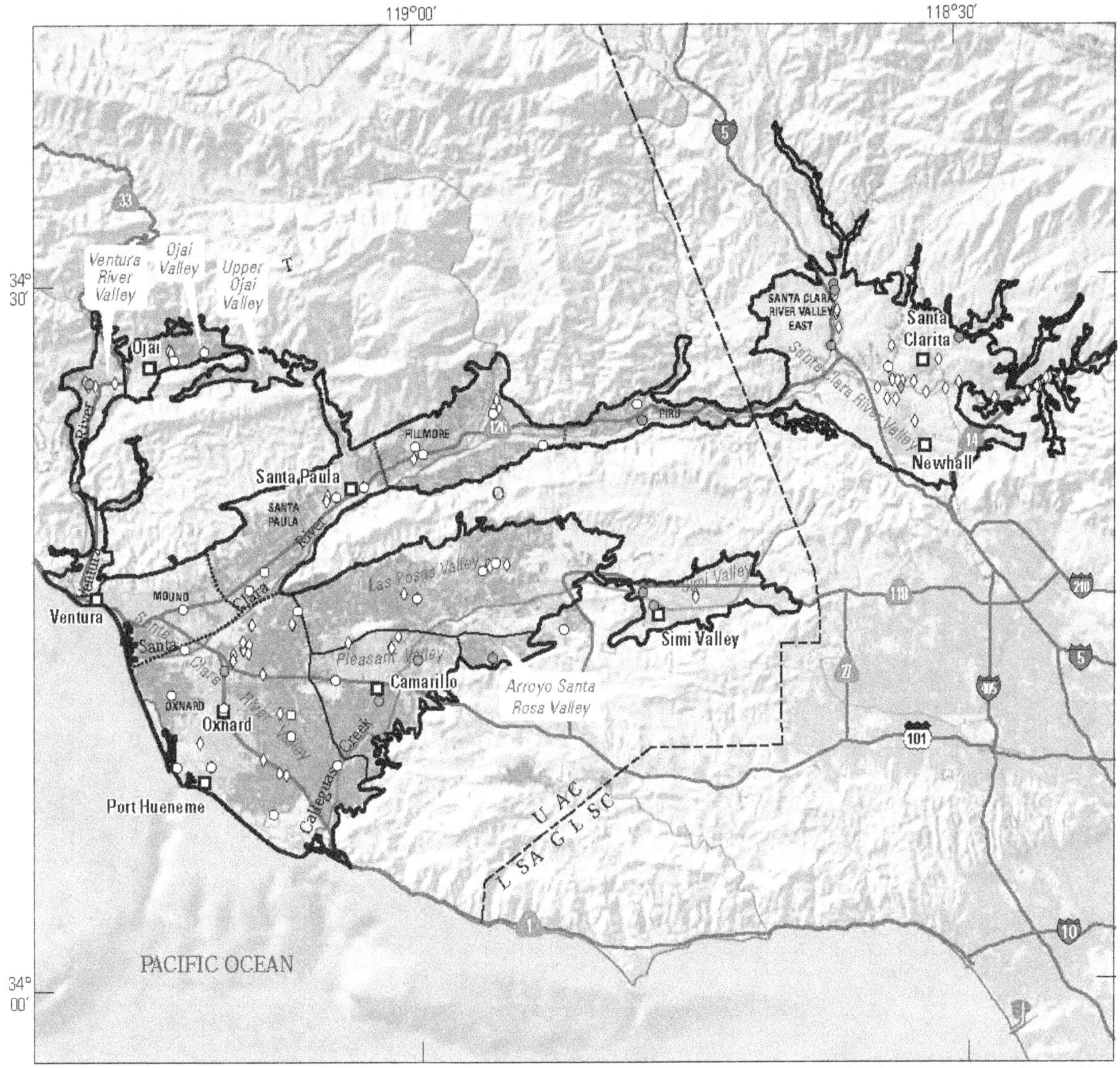

S aded relief derived from U S Geologi al Survey
ational levation Dataset 00
Albers ual Area Coni Pro e tion
ort Ameri an Datum of (AD)

0 5 10 MILES
0 5 10 KILOMETERS

Land-use data from a aga i and ot ers 00

. Pesti ides

EXPLANATION

LAN E
LA I I ATION

Urban

Agri ultural

atural

(from California
De artment of ater esour es 00)

MOUND

N	USGS-grid wells	California De artment of Publi Healt (CDPH-ot er) wells
ot dete ted	○	◊
to esti ides	●	♦
5 to 0 esti ides	●	♦

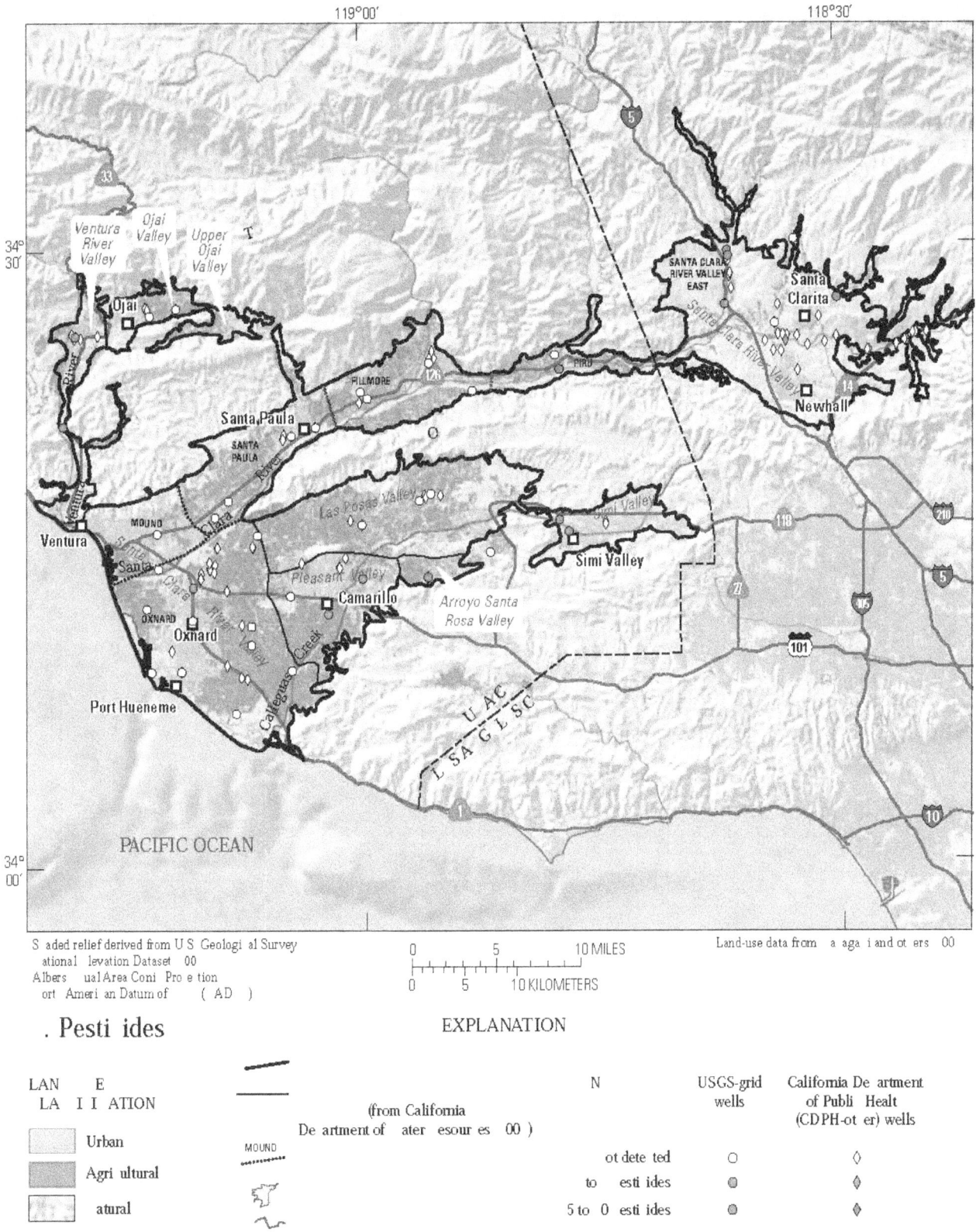

Figure 19. Number of pesticide detections and relative-concentrations of selected organic constituents for USGS-grid wells and CDPH wells, Santa Clara River Valley study unit, California GAMA Priority Basin Project.

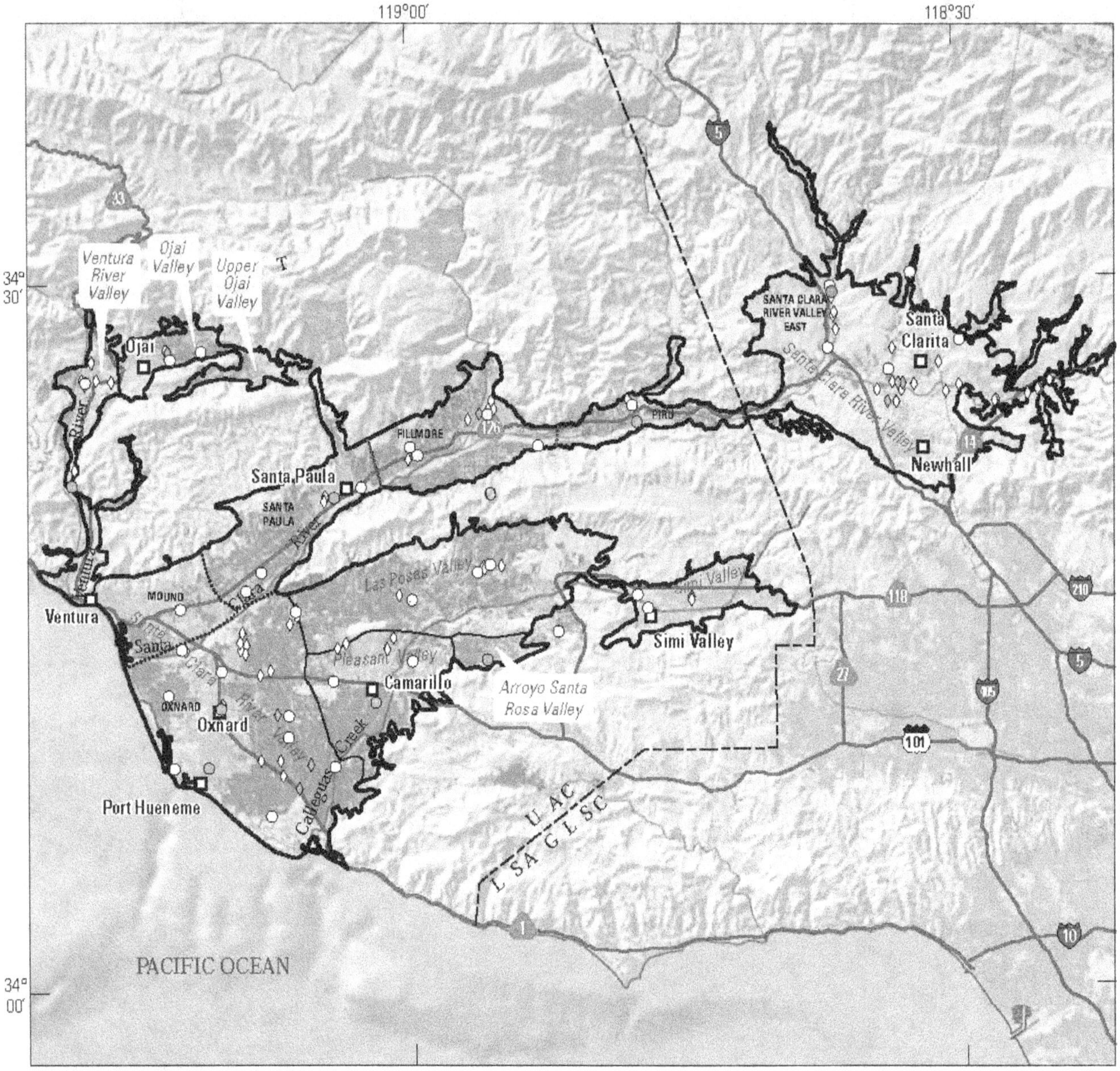

S aded relief derived from U S Geologi al Survey
ational levation Dataset 00
Albers ual Area Coni Pro e tion
ort Ameri an Datum of (AD)

0 5 10 MILES
0 5 10 KILOMETERS

Land-use data from a aga i and ot ers 00

B. ri alomet anes

EXPLANATION

LAN E
LA I I ATION

Urban

Agri ultural

atural

(from California
De artment of ater esour es 00)

MOUND

	USGS-grid wells	California De artment of Publi Healt (CDPH-ot er) wells
ot dete ted	○	◊
Low	○	◊
oderate	○	◊

Figure 19.—Continued

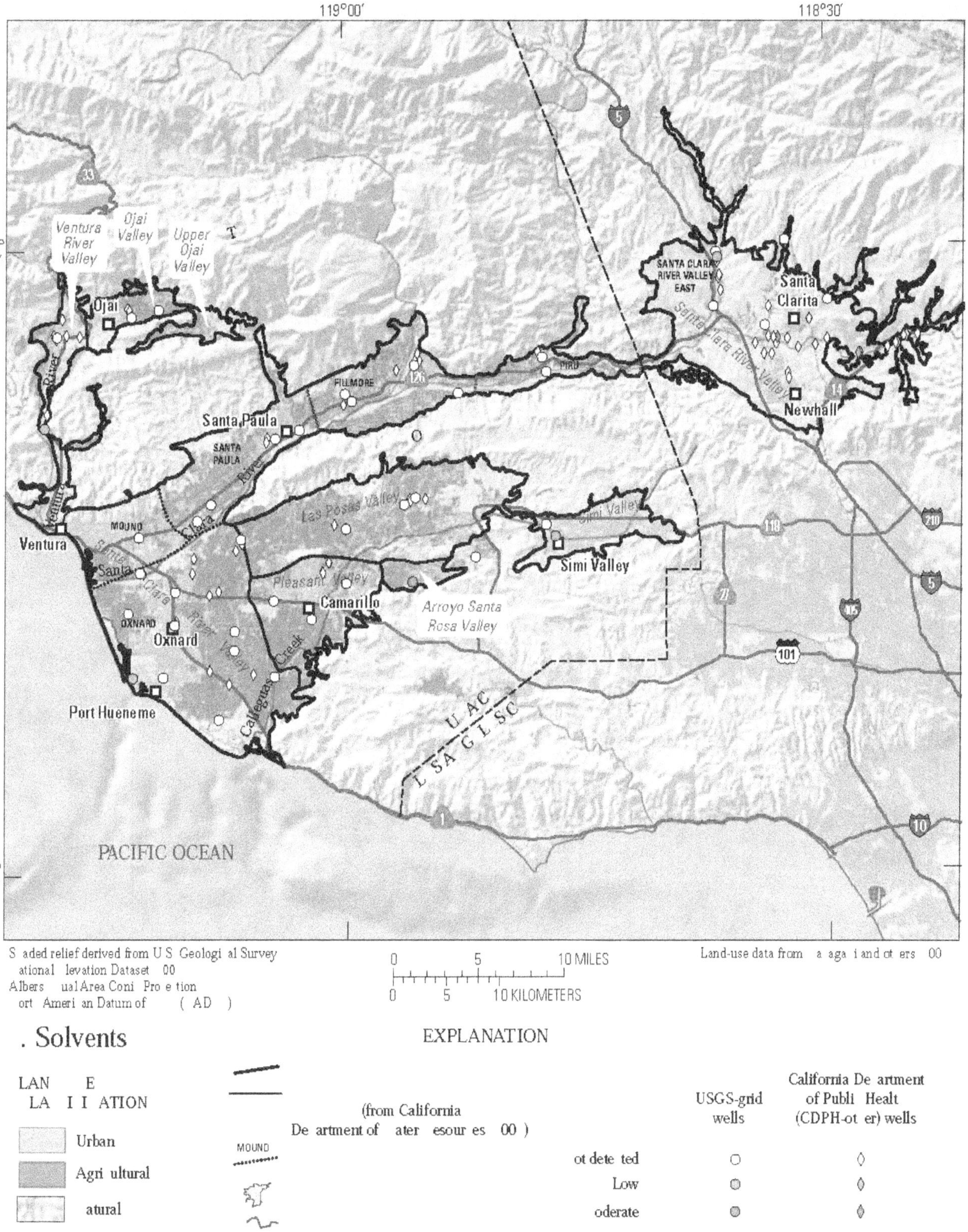

Figure 19.—Continued

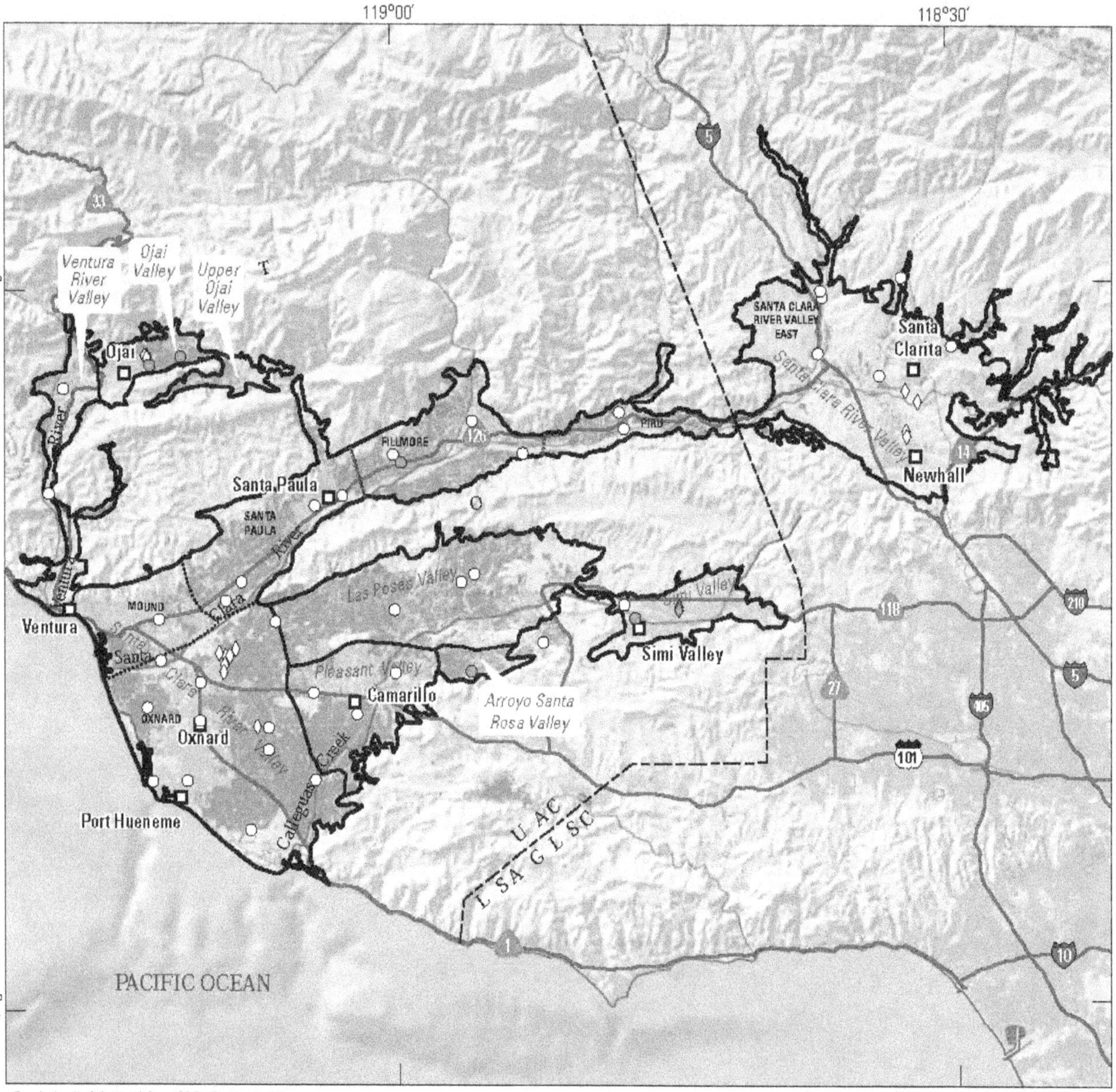

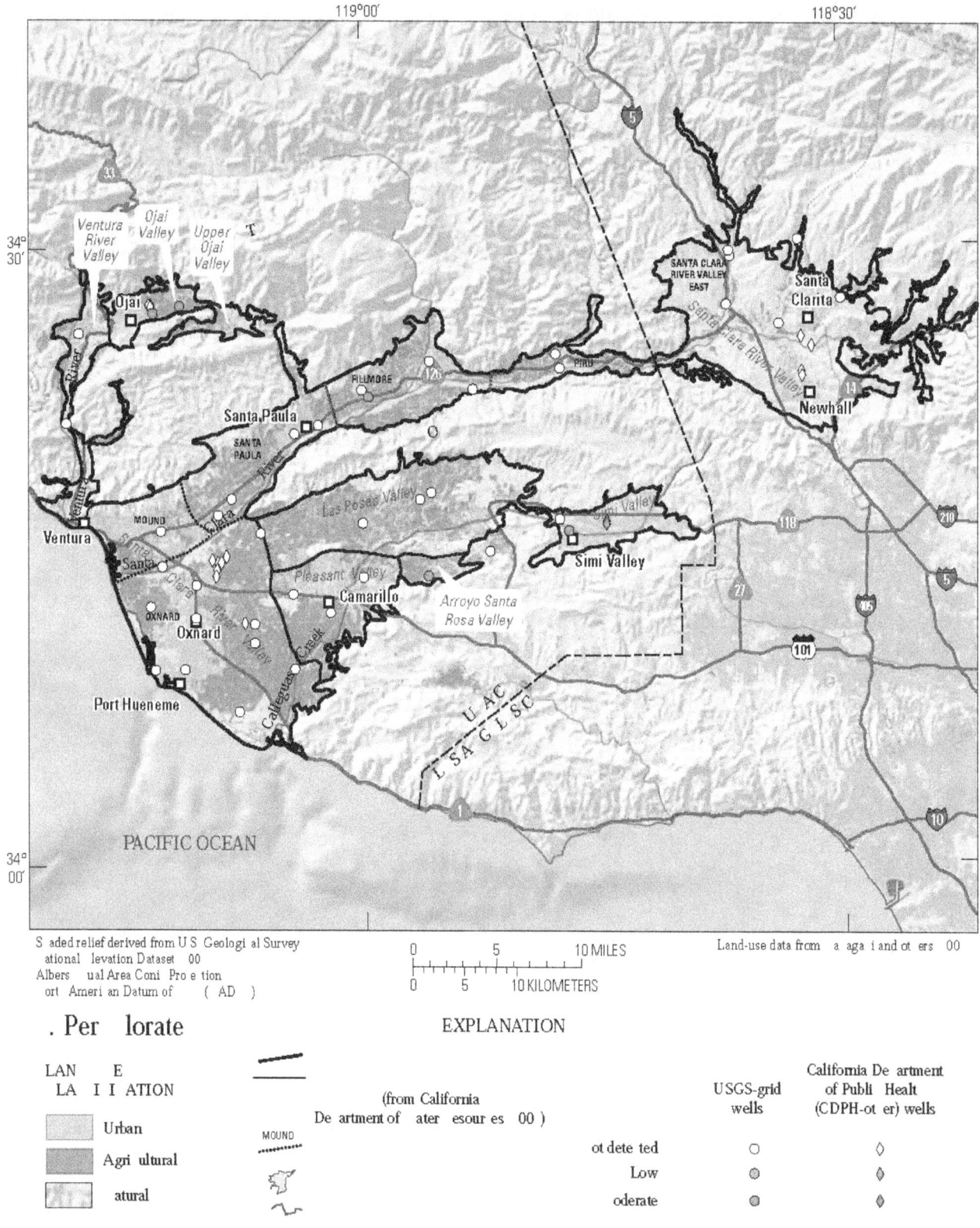

S aded relief derived from U S Geologi al Survey
ational levation Dataset 00
Albers ual Area Coni Pro e tion
ort Ameri an Datum of (AD)

0 5 10 MILES
0 5 10 KILOMETERS

Land-use data from a aga i and ot ers 00

. Per lorate

EXPLANATION

LAN E
LA I I ATION

Urban

Agri ultural

atural

(from California
De artment of ater esour es 00)

MOUND

	USGS-grid wells	California De artment of Publi Healt (CDPH-ot er) wells
ot dete ted	○	◇
Low	●	◆
oderate	●	◆

Figure 19.—Continued

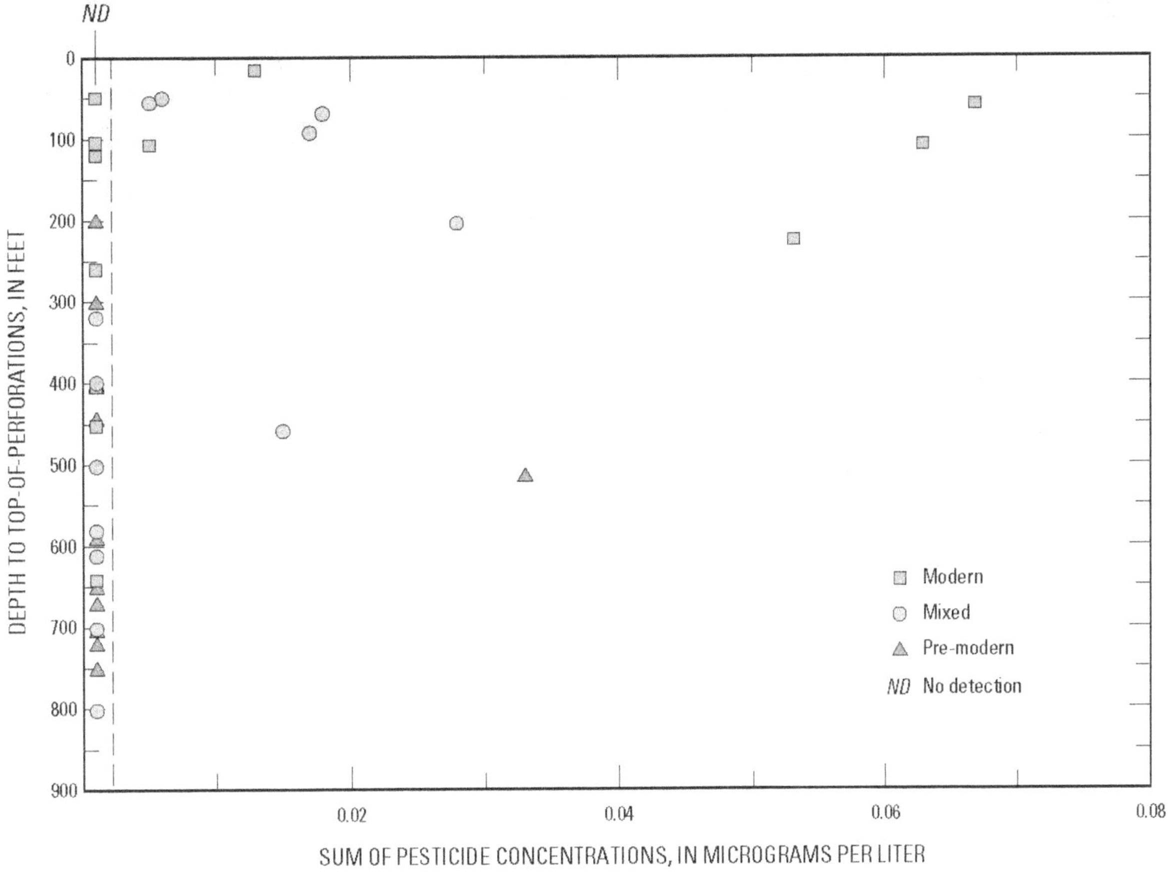

Figure 20. Relation of pesticide concentrations to depth to top-of-perforations, classified by groundwater age, in USGS-grid wells, Santa Clara River Valley study unit, California GAMA Priority Basin Project.

Factors Affecting Perchlorate

Perchlorate was negatively correlated with pH (table 11) but was not correlated with other explanatory factors or groundwater age class (tables 6 and 11). The absence of correlation may be due to the relatively low detection frequency. The negative correlation of perchlorate to pH (table 11) may indicate that perchlorate concentrations decrease with increasing depth, with increasing pH serving as a proxy indicator for increasing depth (table 7, fig. 9*B*). A direct correlation of perchlorate to depth may not have been discernible because well construction data were available for only four of the five wells with detections of perchlorate.

In addition, perchlorate concentrations were positively correlated with nitrate concentrations ($\rho = 0.419$, $\rho = 0.033$; not included in table 11). Nitrate (as nitrogen) concentrations were greater than the MCL-US of 10 mg/L for four of the five wells (SCRV-04, -13, -28, -35, table A1, fig. 19*D*) with perchlorate concentrations greater than the reporting level. The fifth well (SCRV-38) with reported perchlorate greater than the reporting level was not sampled by USGS Priority Basin Project for nutrients, and nitrate values were not reported in the CDPH database. Agricultural land use was the dominant land use around four of the five wells with perchlorate detections, and urban land use was the dominant land use around the other well (SCRV-35). High nitrate concentrations can be associated with several sources, including fertilizers. Possible anthropogenic sources of perchlorate could include industrial, manufacturing, or commercial uses, such as explosives, road flares, automobile air-bag systems, or other products (Parker and others, 2008), or nitrate fertilizers, which were mined from the Atacama Desert of Chile and used historically on some orchard crops (Dasgupta and others, 2006). Elevated concentrations of perchlorate can also result from mobilization of naturally-occurring perchlorate by downward moving irrigation return flows (Fram and Belitz, 2011). The association of perchlorate detections and elevated nitrate in four wells, along with surrounding agricultural land use, is consistent with the mobilization of naturally-occuring perchlorate by irrigation or from application of perchlorate-bearing fertilizer; additional investigation is needed to clarify sources of perchlorate in the study unit.

Summary

Groundwater quality in the approximately 460 mi^2 Santa Clara River Valley study unit (SCRV) was investigated as part of the Priority Basin Project of the Groundwater Ambient Monitoring and Assessment (GAMA) program. Samples were collected by the U.S. Geological Survey (USGS) from April through June 2007.

The GAMA Priority Basin Project is designed to provide a statistically robust characterization of untreated groundwater quality in the primary aquifers at the basin-scale. Forty-two grid wells were randomly selected within spatially distributed grid cells across the SCRV to assess the quality of the groundwater. An additional 11 wells were sampled to improve understanding of the relation of water quality to explanatory factors. Samples from USGS-grid wells were analyzed for as many as 374 constituents. CDPH inorganic data from the prior 3-year period (November 1, 2003, to October 31, 2006) were used to complement USGS-grid well data and to provide additional information about groundwater quality.

Relative-concentrations (sample concentration divided by water-quality benchmark concentration) were used for evaluating groundwater quality. Selected constituents with high (relative-concentration greater than 1) or moderate relative-concentrations or detection frequencies greater than or equal to 10 percent were selected to focus the *understanding assessment* on those constituents with the greatest effect on water quality. The relative-concentration threshold for classifying inorganic constituents as moderate was 0.5 and for classifying organic constituents 0.1; the lower threshold value for organic constituents was selected because these constituents, which are typically from anthropogenic sources, have smaller relative-concentrations and generally are less prevalent than naturally occurring inorganic constituents.

Aquifer-scale proportion was used as a metric for assessing the quality of untreated groundwater for the SCRV. Aquifer-scale proportions are defined as the percentage of the area of the primary aquifer system with concentrations greater than or less than specified thresholds relative to regulatory or aesthetic benchmarks. Grid-based and spatially weighted statistical approaches were used to assess aquifer-scale proportions of constituents at high, moderate, and low relative-concentrations in the primary aquifers. The grid-based approach was used for most constituents unless the spatially weighted proportions were significantly different. Raw detection frequencies also were calculated by using all available data and are provided for comparison but are not used in the analysis because detection frequencies are potentially biased because the wells are not uniformly distributed.

For inorganic constituents with human-health benchmarks, the relative-concentrations of at least one constituent were high in 21 percent of the primary aquifer system, and were moderate in 30 percent of the primary aquifer system. The inorganic constituents with human-health benchmarks and the largest aquifer-scale high proportions were nitrate (15 percent), gross alpha radioactivity (14 percent), vanadium (3.4 percent), boron (3.2 percent), and arsenic (2.3 percent).

For inorganic constituents with aesthetic benchmarks (SMCL), the relative-concentrations of at least one constituent were high in 54 percent of the primary aquifer system, and were moderate in 41 percent of the primary aquifer system. Inorganic constituents with SMCLs and high relative-concentrations included TDS (35 percent), sulfate (22 percent), manganese (38 percent), and iron (22 percent).

Of the 88 VOCs and gasoline oxygenates analyzed, 17 were detected. Of the 17 VOCs detected, 15 had human-health benchmarks. None of the VOCs were detected at high relative-concentrations. The relative-concentrations of two solvents, carbon tetrachloride and TCE, were moderate in 2.4 percent of the primary aquifer system. The relative-concentrations of total THMs also were moderate in 2.4 percent of the primary aquifer system. The remaining VOCs that were detected were detected at low relative-concentrations. Of the 118 pesticides analyzed, 13 were detected. Five of these 13 had human-health benchmarks. Relative-concentrations for pesticides were neither high nor moderate in any part of the primary aquifer system. Two pesticides, simazine and atrazine, were detected at low relative-concentrations in 26 percent and 17 percent, respectively, of the primary aquifer system. Perchlorate was detected at moderate relative-concentrations in 12 percent of the primary aquifer system.

Water quality primarily varied in relation to depth, groundwater age, reduction-oxidation conditions, and pH. Concentration of nitrate and detections of pesticides were high in wells with shallow perforation depths and were correlated with high dissolved-oxygen concentrations. Nitrate concentrations were higher in groundwater with modern and mixed ages than in groundwater with pre-modern age. Decreases in concentrations of total dissolved solids and sulfate with increasing pH probably indicates correlations between these constituents and increasing depth across most of the SCRV. Increases in pH with depth likely are the result of dissolution of primary silicate minerals in the aquifer. Manganese and iron concentrations were largest in deeper wells, in pre-modern age groundwater, and in the downgradient part of the SCRV (closest to the coastline), indicating the prevalence of reducing oxidation-reduction conditions in these aquifer zones.

Acknowledgments

The authors thank the following cooperators for their support: the California State Water Board, Lawrence Livermore National Laboratory, California Department of Public Health, and California Department of Water Resources. We especially thank the well owners and water purveyors for their cooperation in allowing the U.S. Geological Survey to collect samples from their wells. Funding for this work was provided by State of California bonds authorized by Proposition 50 and administered by the State Water Board.

References Cited

Aeschbach-Hertig, W., Peeters, F., Beyerle, U., and Kipfer, R., 1999, Interpretation of dissolved atmospheric noble gases in natural waters: Water Resources Research, v. 35, no. 9, p. 2779–2792.

Aeschbach-Hertig, W., Peeters, F., Beyerle, U., and Kipfer, R., 2000, Paleotemperature reconstruction from noble gases in ground water taking into account equilibration with entrapped air: Nature, v. 405, p. 1040–1044.

Andrews, J.N., 1985, The isotopic composition of radiogenic helium and its use to study groundwater movement in confined aquifers: Chemical Geology, v. 49, p. 339–351.

Andrews, J.N., and Lee, D.J., 1979, Inert gases in groundwater from the Bunter Sandstone of England as indicators of age and paleoclimatic trends: Journal of Hydrology, v. 41, p. 233–252.

Belitz, K., Dubrovsky, N.M., Burow, K.R., Jurgens, B., and Johnson, T., 2003, Framework for a ground-water quality monitoring and assessment program for California: U.S. Geological Survey Water-Resources Investigations Report 03-4166, 78 p.

Belitz, K., Jurgens, B., Landon, M.K., Fram, M.S., and Johnson, T., 2010, Estimation of aquifer scale proportion using equal area grids: Assessment of regional scale groundwater quality: Water Resources Research, v. 46, W11550, 14 p.

Bennett, G.L., V, Fram, M.S., and Belitz, Kenneth, 2011, Status and understanding of groundwater quality in the Southern, Middle, and Northern Sacramento Valley study units, 2005–08—California GAMA Priority Basin Project: U.S. Geological Survey Scientific Investigations Report 2011–5002, 86 p.

Bennett, G.L., V, Fram, M.S., Belitz, Kenneth, and Jurgens, B.C., 2010, Status and understanding of groundwater quality in the Northern San Joaquin Basin, 2005—California GAMA Priority Basin Project: U.S. Geological Survey Scientific Investigations Report 2010–5175, 82 p.

Brown, L.D., Cai, T.T., and Dasgupta, A., 2001, Interval estimation for a binomial proportion: Statistical Science, v. 16, no. 2, p. 101–117.

Burow, K.R., Dubrovsky, N.M., and Shelton, J.L., 2007, Temporal trends in concentrations of DBCP and nitrate in groundwater in the eastern San Joaquin Valley, California, USA: Hydrogeology Journal, v. 15, no. 5, p. 991–1007.

California Department of Health Services, 2007, California Code of Regulations, Title 22, Social Security, Division 4, Environmental Health, Chapter 15 Domestic Water Quality and Monitoring Regulations, Articles 4, 5.5, and 16, Register 2007, no. 4, accessed February 10, 2007, at http://ccr.oal.ca.gov/.

California Department of Public Health, 2008a, California drinking water-related laws: California Department of Public Health website, accessed June 1, 2008, at http://www.cdph.ca.gov/certlic/drinkingwater/Pages/Lawbook.aspx.

California Department of Public Health, 2008b, Perchlorate in drinking water: California Department of Public Health website, accessed January 2, 2008, at http://www.cdph.ca.gov/certlic/drinkingwater/Pages/Perchlorate.aspx.

California Department of Water Resources, 1965, Seawater intrusion, Oxnard Plain of Ventura County, Bulletin No. 63-1 [variously paged].

California Department of Water Resources, 2001, Land use for Ventura County, California, for 1995 [digital data]: California Department of Water Resources, Division of Planning, Statewide Planning Branch, Land and Water Use website accessed November 15, 2010, at http://www.water.ca.gov/landwateruse/lusrvymain.cfm.

California Department of Water Resources, 2003, California's groundwater: California Department of Water Resources Bulletin 118, 246 p., accessed September 27, 2010, at http://www.dpla2.water.ca.gov/publications/groundwater/bulletin118/Bulletin118_Entire.pdf.

California Department of Water Resources, 2004a, Arroyo Santa Rosa Valley groundwater basin: California's Groundwater Bulletin 118, accessed August 22, 2007, at http://www.dpla2.water.ca.gov/publications/groundwater/bulletin118/basins/pdfs_desc/4-7.pdf.

California Department of Water Resources, 2004b, Santa Clara River Valley basin, Fillmore subbasin: California's Groundwater Bulletin 118, accessed August 22, 2007, at http://www.dpla2.water.ca.gov/publications/groundwater/bulletin118/basins/pdfs_desc/4-4.05.pdf.

California Department of Water Resources, 2004c, Las Posas Valley groundwater basin: California's Groundwater Bulletin 118, accessed August 22, 2007, at http://www.dpla2.water.ca.gov/publications/groundwater/bulletin118/basins/pdfs_desc/4-8.pdf.

California Department of Water Resources, 2004d, Ventura River Valley groundwater basin, Lower Ventura River subbasin: California's Groundwater Bulletin 118, accessed August 22, 2007, at http://www.dpla2.water.ca.gov/publications/groundwater/bulletin118/basins/pdfs_desc/4-3.02.pdf.

California Department of Water Resources, 2004e, Santa Clara River Valley basin, Mound subbasin: California's Groundwater Bulletin 118, accessed August 22, 2007, at http://www.dpla2.water.ca.gov/publications/groundwater/bulletin118/basins/pdfs_desc/4-4.03.pdf.

California Department of Water Resources, 2004f, Ojai Valley groundwater basin: California's Groundwater Bulletin 118, accessed August 22, 2007, at http://www.dpla2.water.ca.gov/publications/groundwater/bulletin118/basins/pdfs_desc/4-2.pdf.

California Department of Water Resources, 2004g, Santa Clara River Valley groundwater basin, Oxnard subbasin: California's Groundwater Bulletin 118, accessed August 22, 2007, at http://www.dpla2.water.ca.gov/publications/groundwater/bulletin118/basins/pdfs_desc/4-4.02.pdf.

California Department of Water Resources, 2004h, Santa Clara River Valley basin, Piru subbasin: California's Groundwater Bulletin 118, accessed August 22, 2007, at http://www.dpla2.water.ca.gov/publications/groundwater/bulletin118/basins/pdfs_desc/4-4.06.pdf.

California Department of Water Resources, 2004i, Pleasant Valley groundwater basin: California's Groundwater Bulletin 118, accessed August 22, 2007, at http://www.dpla2.water.ca.gov/publications/groundwater/bulletin118/basins/pdfs_desc/4-6.pdf.

California Department of Water Resources, 2004j, Santa Clara River Valley basin, Santa Paula subbasin: California's Groundwater Bulletin 118, accessed August 22, 2007, at http://www.dpla2.water.ca.gov/publications/groundwater/bulletin118/basins/pdfs_desc/4-4.04.pdf.

California Department of Water Resources, 2004k, Santa Clara River Valley groundwater basin, Santa Clara River Valley East subbasin: California's Groundwater Bulletin 118, accessed August 22, 2007, at http://www.dpla2.water.ca.gov/publications/groundwater/bulletin118/basins/pdfs_desc/4-4.07.pdf.

California Department of Water Resources, 2004l, Simi Valley groundwater basin: California's Groundwater Bulletin 118, accessed August 22, 2007, at http://www.dpla2.water.ca.gov/publications/groundwater/bulletin118/basins/pdfs_desc/4-9.pdf.

California Department of Water Resources, 2004m, Upper Ojai Valley groundwater basin: California's Groundwater Bulletin 118, accessed August 22, 2007, at http://www.dpla2.water.ca.gov/publications/groundwater/bulletin118/basins/pdfs_desc/4-1.pdf.

California Department of Water Resources, 2004n, Ventura River Valley groundwater basin, Upper Ventura River subbasin: California's Groundwater Bulletin 118, accessed August 22, 2007, at http://www.dpla2.water.ca.gov/publications/groundwater/bulletin118/basins/pdfs_desc/4-3.01.pdf.

California Environmental Protection Agency, 2010, GAMA—Groundwater Ambient Monitoring and Assessment Program: State Water Resources Control Board website, accessed September 9, 2010, at http://www.waterboards.ca.gov/water_issues/programs/gama.

California State Water Resources Control Board, 2003, A comprehensive groundwater quality monitoring program for California: Assembly Bill 599 Report to the Governor and Legislature, March 2003, 100 p., accessed January 19, 2011, at http://www.waterboards.ca.gov/water_issues/programs/gama/docs/final_ab_599_rpt_to_legis_7_31_03.pdf.

Cey, B.D., Hudson, G.B., Moran, J.E., and Scanlon, B.R., 2008, Impact of artificial recharge on dissolved noble gases in groundwater in California: Environmental Science Technology, v. 42, p. 1017–1023.

Chapelle, F.H., 2001, Ground-water microbiology and geochemistry (2d ed.): New York, John Wiley and Sons, Inc., 477 p.

Chapelle, F.H., McMahon, P.B., Dubrovsky, N.M., Fuji, R.F., Oaksford, E.T., and Vroblesky, D.A., 1995, Deducing the distribution of terminal electron-accepting processes in hydrologically diverse groundwater systems: Water Resources Research, v. 31, no. 2, p. 359–371.

Clark, I.D., and Fritz, P., 1997, Environmental isotopes in hydrogeology: New York, Lewis Publishers, 328 p.

Craig, H., and Lal, D., 1961, The production rate of natural tritium: Tellus, v. 13, p. 85–105.

Dahlen, M.Z., Osborne, R.H., and Gorsline, D.S., 1990, Late Quaternary history of the Ventura mainland shelf, California: Marine Geology, v. 94, p. 317–340.

Dasgupta, P.K., Dyke, J.V., Kirk, A.B., and Jackson, W.A., 2006, Perchlorate in the United States—Analysis of relative source contributions to the food chain: Environmental Science and Technology, v. 40, p. 6608–6614.

Davis, G.H., 1961, Geologic control of mineral composition of stream waters of the eastern slope of the Southern Coast Ranges California: U.S. Geological Survey Water-Supply Paper 1535-B, 24 p.

Davis, S., and DeWiest, R.J., 1966, Hydrogeology: New York, John Wiley and Sons, 413 p.

Dibblee, T.W., 1988, Geologic map of the Ventura/Pitas Point Quadrangles, Ventura County, California: Santa Barbara, Calif., Dibblee Geological Foundation, Dibblee Foundation Map series, DF-21, scale 1:24,000.

Dibblee, T.W., 1990a, Geologic map of the Fillmore Quadrangle, Ventura County, California: Santa Barbara, Calif., Dibblee Geological Foundation, Dibblee Foundation Map series, DF-27, scale 1:24,000.

Dibblee, T.W., 1990b, Geologic map of the Santa Paula Peak Quadrangle, Ventura County, California: Santa Barbara, Calif., Dibblee Geological Foundation, Dibblee Foundation Map series, DF-26, scale 1:24,000.

Dibblee, T.W., 1991, Geologic map of the Piru Quadrangle, Ventura County, California: Santa Barbara, Calif., Dibblee Geological Foundation, Dibblee Foundation Map series, DF-34, scale 1:24,000.

Dibblee, T.W., 1992a, Geologic map of the Simi Quadrangle, Ventura County, California: Santa Barbara, Calif., Dibblee Geological Foundation, Dibblee Foundation Map series, DF-39, scale 1:24,000.

Dibblee, T.W., 1992b, Geologic map of the Moorpark Quadrangle, Ventura County, California: Santa Barbara, Calif., Dibblee Geological Foundation, Dibblee Foundation Map series, DF-40, scale 1:24,000.

Dibblee, T.W., 1992c, Geologic map of the Santa Paula quadrangle, Ventura County, California: Santa Barbara, Calif., Dibblee Geological Foundation, Dibblee Foundation Map series, DF-41, scale 1:24,000.

Dibblee, T.W., 1992d, Geologic map of the Saticoy quadrangle, Ventura County, California: Santa Barbara, Calif., Dibblee Geological Foundation, Dibblee Foundation Map series, DF-42, scale 1:24,000.

Dotsika, E., Poutoukis, D., Michelot, J.L., and Kloppmann, W., 2005, Stable isotope and chloride, boron study for tracing sources of boron contamination in groundwater—Boron contents in fresh and thermal water in different areas of Greece: Water, Air, and Soil Pollution, v. 174, p. 19–32.

Emery, K.O., 1960, The Sea off Southern California: New York, J. Wiley & Sons Inc., p. 366.

Fontes, J.C., and Garnier, J.M., 1979, Determination of the initial ^{14}C activity of the total dissolved carbon—A review of the existing models and a new approach: Water Resources Research, v. 15, p. 399–413.

Fox Canyon Groundwater Management Agency, 2007, 2007 Update to the Fox Canyon Groundwater Management Agency Groundwater Water Management Plan May 2007: Ventura, Calif., Fox Canyon Groundwater Management Agency, 128 p., accessed September 27, 2010, at http://www.fcgma.org/publicdocuments/plans.shtml.

Fram, M.S., and Belitz, Kenneth, 2011, Probability of detecting perchlorate under natural conditions in deep groundwater in California and the southwestern United States: Environmental Science and Technology, v. 45, p. 1271–1277.

Gilliom, R.J., Barbash, J.E., Crawford, C.G., Hamilton, P.A., Martin, J.D., Nakagaki, N., Nowell, L.H., Scott, J.C., Stackelberg, P.E., Thelin, G.P., and Wolock, D.M., 2006, The quality of our nation's waters—Pesticides in the nation's streams and ground water, 1992–2001: U.S. Geological Survey Circular 1291, 172 p.

Goodarzi, F., and Swaine, D.J., 1994, The influence of geologic factors on the concentration of boron in Australian and Canadian coals: Chemical Geology, v. 118, p. 301–318.

Greene, G.H., Wolf, S.C., and Blom, K.G., 1978, The marine geology of the eastern Santa Barbara Channel with particular emphasis on the ground water basins offshore from Oxnard Plain, Southern California: U.S. Geological Survey Open-File Report 78-305, 104 p.

Gutierrez-Alonso, G., and Gross, M.R., 1997, Geometry of inverted faults and related folds in the Monterey Formation—Implications for the structural evolution of the southern Santa Maria basin, California: Journal of Structural Geology, v. 19, no. 10, p. 1303–1321.

Hallberg, G.R., and Keeney, D.R., 1993, Nitrate, *in* Alley, W.M., ed., Regional ground-water quality: New York, Van Nostrand Reinhold, p. 297–322.

Hanson, R.T., Martin, Peter, and Koczot, K.M., 2003, Simulation of ground-water/surface-water flow in the Santa Clara–Calleguas ground-water basin, Ventura County, California: U.S. Geological Survey Water-Resources Investigations Report 02-4136, 157 p.

Helsel, D.R., and Hirsch, R.M., 2002, Statistical methods in water resources: U.S. Geological Survey Techniques of Water-Resources Investigations, book 4, chap. A3, 510 p. (Also available at http://water.usgs.gov/pubs/twri/twri4a3/.)

Hem, J.D., 1985, Study and interpretation of the chemical characteristics of natural water (3d ed.): U.S. Geological Survey Water-Supply Paper 2254, 263 p.

Hsü, K.J., Kelts, K., and Valentine, J.W., 1980, Resedimented facies in Ventura Basin, California, and model of longitudinal transport of turbidity currents: The American Association of Petroleum Geologists Bulletin, v. 64, no. 7, p. 1034–1051.

Isaaks, E.H., and Srivastava, R.M., 1989, Applied Geostatistics: New York, Oxford University Press, 561 p.

Izbicki, J.A., 1996a, Seawater intrusion in a coastal California aquifer: U.S. Geological Survey Fact Sheet 125-96, 4 p.

Izbicki, J.A., 1996b, Source, movement, and age of ground water in a coastal California aquifer: U.S. Geological Survey Fact Sheet 126-96, 4 p.

Izbicki, J.A., 1996c, Use of $\delta^{18}O$ and δD to define seawater intrusion, *in* Bathala, C.T., ed., Proceedings of the North American Ground Water and Environment Congress, June 23–28, 1996, Anaheim, Calif.: New York, American Society of Civil Engineers, p. 4306–4311.

Izbicki, J.A., 2004, A small-diameter sample pump for collection of depth-dependent samples from production wells under pumping conditions: U.S. Geological Survey Fact Sheet 2004-3096, 2 p.

Izbicki, J.A., Christensen, A.H., Newhouse, M.N., and Aiken, G.R., 2005, Inorganic, isotopic, and organic composition of high-chloride water from wells in a coastal southern California aquifer: Applied Geochemistry, v. 20, p. 1496–1517.

Izbicki, J.A., and Martin, Peter, 1997, Use of isotopic data to evaluate recharge and geologic controls on the movement of ground water in Las Posas Valley, Ventura County, California: U.S. Geological Survey Water-Resources Investigations Report 97-4035, 12 p.

Izbicki, J.A., Martin, Peter, Densmore, J.N., and Clark, D.A., 1995, Water-quality data for the Santa Clara–Calleguas Hydrologic Unit, Ventura County, California, October 1989 through December 1993: U.S. Geological Survey Open-File Report 95-315, 124 p.

Jennings, C.W., 1977, Geologic map of California: California Department of Conservation, Division of Mines and Geology, Geologic Data Map No. 2, scale 1:750,000.

Johnson, T.D., and Belitz, K., 2009, Assigning land use to supply wells for the statistical characterization of regional groundwater quality—Correlating urban land-use and VOC occurrence: Journal of Hydrology, v. 370, p. 100–108.

Jurgens, B.C., Fram, M.S., Belitz, Kenneth, Burow, K.R., and Landon, M.K., 2010, Effects of ground-water development on uranium—Central Valley, California, USA: Ground Water, v. 48, p. 913–928.

Jurgens, B.C., McMahon, P.B., Chapelle, F.H., and Eberts, S.M., 2009, An Excel® workbook for identifying redox processes in ground water: U.S. Geological Survey Open-File Report 2009–1004, 8 p. (Also available at http://pubs.usgs.gov/of/2009/1004/.)

Kalin, R.M., 2000, Radiocarbon dating of groundwater systems, *in* Cook, P.G., and Herczeg, A., eds., Environmental tracers in subsurface hydrology: Boston, Kluwer Academic Publishers, p. 111–144.

Kendall, C., 1998, Tracing nitrogen sources and cycling in catchments, *in* Kendall, C., and McDonnell, J.J., eds., Isotope tracers in catchment hydrology: Amsterdam, Elsevier Science, chap. 16, p. 519–576.

Kulongoski, J.T., Belitz, Kenneth, Landon, M.K., and Farrar, Christopher, 2010, Status and understanding of groundwater quality in the North San Francisco Bay groundwater basins, 2004—California GAMA Priority Basin Project: U.S. Geological Survey Scientific Investigations Report 2010-5089, 88 p.

Kulongoski, J.T., Hilton, D.R., Cresswell, R.G., Hostetler, S., and Jacobson, G., 2008, Helium-4 characteristics of groundwaters from Central Australia—Comparative chronology with chlorine-36 and carbon-14 dating techniques: Journal of Hydrology, v. 348, p. 176–194.

Landon, M.K., Belitz, Kenneth, Jurgens, B.C., Kulongoski, J.T., and Johnson, T., 2010, California GAMA Program—Status and understanding of ground-water quality in the Central-Eastside San Joaquin Basin, 2006—U.S. Geological Survey Scientific Investigations Report 2009-5266, 97 p.

Lindburg, R.D., and Runnells, D.D., 1984, Groundwater redox reactions: Science, v. 225, p. 925–927.

Lucas, L.L., and Unterweger, M.P., 2000, Comprehensive review and critical evaluation of the half-life of tritium: Journal of Research of the National Institute of Standards and Technology, v. 105, no. 4, p. 541–549.

Mangelodorf, K., and Rullkötter, J., 2003, Natural supply of oil-derived hydrocarbons into marine sediments along California continental margin during the late Quaternary: Organic Geochemistry, v. 34, p. 1145–1159.

Manning, A.H., Solomon, D.K., and Thiros, S.A., 2005, $^3H/^3He$ age data in assessing the susceptibility of wells to contamination: Ground Water, v. 43, no. 3, p. 353–367.

McMahon, P.B., and Chapelle, F.H., 2008, Redox processes and water quality of selected principal aquifer systems: Ground Water, v. 46, no. 2, p. 259–271.

Michel, R.L., 1989, Tritium deposition in the continental United States, 1953–83: U.S. Geological Survey Water-Resources Investigations Report 89-4072, 46 p.

Michel, R.L., and Schroeder, R., 1994, Use of long-term tritium records from the Colorado River to determine timescales for hydrologic processes associated with irrigation in the Imperial Valley, California: Applied Geochemistry, v. 9, p. 387–401.

Montrella, Joseph, and Belitz, Kenneth, 2009, Ground-water quality data in the Santa Clara River Valley study unit, 2007—Results from the California GAMA Program: U.S. Geological Survey Data Series 408, 84 p. (Also available at http://pubs.usgs.gov/ds/408.)

Moore, K.B., Ekwurzel, B., Esser, B.K., Hudson, G.B., and Moran, J., 2006, Sources of groundwater nitrate revealed using residence time and isotope methods: Applied Geochemistry, v. 21, p. 1016–1029.

Morrison, P., and Pine, J., 1955, Radiogenic origin of the helium isotopes in rock: Annals of the New York Academy of Sciences, v. 12, p. 19–92.

Nakagaki, Naomi, Price, C.V., Falcone, J.A., Hitt, K.J., and Ruddy, B.C., 2007, Enhanced National Land Cover Data 1992 (NLCDe 92): U.S. Geological Survey Raster digital data, accessed September 27, 2010, at http://water.usgs.gov/lookup/getspatial?nlcde92.

Nakagaki, Naomi, and Wolock, D.M., 2005, Estimation of agricultural pesticide use in drainage basins using land cover maps and county pesticide data: U.S. Geological Survey Open-File Report 2005-1188.

Nolan, B.T., and Hitt, K.J., 2006, Vulnerability of shallow groundwater and drinking-water wells to nitrate in the United States: Environmental Science and Technology, v. 40, no. 24, p. 7834–7840.

Parker, D.R., Seyfferth, A.L., and Kiel Reese, B., 2008, Perchlorate in groundwater—A synoptic survey of "pristine" sites in the coterminous United States: Environmental Science and Technology, v. 42, no. 5, p. 1465–1471.

Piper, A.M., 1944, A graphic procedure in the geochemical interpretation of water analyses: American Geophysical Union Transactions, v. 25, p. 914–923.

Plummer, L.N., Michel, R.L., Thurman, E.M., and Glynn, P.D., 1993, Environmental tracers for age-dating young ground water, *in* Alley, W.M., ed., Regional ground-water quality: New York, Van Nostrand Reinhold, p. 255–294.

Poreda, R.J., Cerling, T.E., and Salomon, D.K., 1988, Tritium and helium isotopes as hydrologic tracers in a shallow unconfined aquifer: Journal of Hydrology, v. 103, p. 1–9.

Reichard, E.G., Crawford, S.M., Paybins, K.S., Martin, Peter, Land, Michael, and Nishikawa, Tracy, 1999, Evaluation of surface-water/ground-water interactions in the Santa Clara River Valley, Ventura County, California: U.S. Geological Survey Water-Resources Investigations Report 98-4208, 58 p.

Reimann, C., and de Caritat, P., 1998, Chemical elements in the environment. Factsheets for the Geochemist and Environmental Scientist: Berlin, Springer-Verlag, 398 p.

Rowe, B.L., Toccalino, P.L., Moran, M.J., Zogorski, J.S., and Price, C.V., 2007, Occurrence and potential human-health relevance of volatile organic compounds in drinking water from domestic wells in the United States: Environmental Health Perspectives, v. 115, no. 11, p. 1539–1546.

Saucedo, G.J., Bedford, D.R., Raines, G.L., Miller, R.J., and Wentworth, C.M., 2000, GIS data for the geologic map of California (version 2.0): Sacramento, Calif., California Department of Conservation, Division of Mines and Geology.

Schlosser, P., Stute, M., Dörr, H., Sonntag, C., and Munnich, K.O., 1988, Tritium/^3He dating of shallow groundwater: Earth and Planetary Science Letters, v. 89, p. 353–362.

Scott, J.C., 1990, Computerized stratified random site selection approaches for design of a ground-water quality sampling network: U.S. Geological Survey Water-Resources Investigations Report 90-4101, 109 p.

Solomon, D.K., and Cook, P.G., 2000, ^3H and ^3He, *in* Cook, P.G. and Herczeg, A.L., eds., Environmental tracers in subsurface hydrology: Boston, Kluwer Academic Press, p. 397–424.

Stankiewicz, B.A., Kruge, M.A., Mastalerz, M., and Salmon, G.L., 1996, Geochemistry of the alginite and amorphous organic matter from Type II-S kerogens: Organic Geochemistry, v. 24, no. 5, p. 495–509.

State of California, 1999, Supplemental Report of the 1999 Budget Act 1999–00 Fiscal Year, Item 3940-001-0001, State Water Resources Control Board, accessed September 9, 2010, at http://www.lao.ca.gov/1999/99-00_supp_rpt_lang.html#3940.

State of California, 2001a, Assembly Bill No. 599, Chapter 522, accessed September 9, 2010, at http://www.swrcb.ca.gov/gama/docs/ab_599_bill_20011005_chaptered.pdf.

State of California, 2001b, Groundwater Monitoring Act of 2001: California Water Code, part 2.76, Sections 10780–10782.3, accessed September 9, 2010, at http://www.leginfo.ca.gov/cgi-bin/displaycode?section=wat&group=10001-11000&file=10780-10782.3.

Takaoka, N., and Mizutani, Y., 1987, Tritiogenic ^{3}He in groundwater in Takaoka: Earth and Planetary Science Letters, v. 85, p. 74–78.

Toccalino, P.L., and Norman, J.E., 2006, Health-based screening levels to evaluate U.S. Geological Survey ground water quality data: Risk Analysis, v. 26, no. 5, p. 1339–1348.

Toccalino, P.L., Norman, J.E., Phillips, R.H., Kauffman, L.J., Stackelberg, P.E., Nowell, L.H., Krietzman, S.J., and Post, G.B., 2004, Application of health-based screening levels to ground-water quality data in a state-scale pilot effort: U.S. Geological Survey Scientific Investigations Report 2004-5174, 14 p.

Tolstikhin, I.N., and Kamenskiy, I.L., 1969, Determination of groundwater ages by the T-3He method. Geochemistry International, v. 6, p. 310–811.

Torgersen, T., 1980, Controls on pore-fluid concentrations of ^{4}He and ^{222}Rn and the calculation of ^{4}He/^{222}Rn ages: Journal of Geochemical Exploration, v. 13, p. 7–75.

Torgersen, T., and Clarke, W.B., 1985, Helium accumulation in groundwater—I. An evaluation of sources and continental flux of crustal ^{4}He in the Great Artesian basin, Australia: Geochimica et Cosmochimica Acta, v. 49, p. 1211–1218.

Torgersen, T., Clarke, W.B., and Jenkins, W.J., 1979, The tritium/helium3 method in hydrology: International Atomic Energy Agency, IAEA-SM-228, v. 49, p. 917–930.

Turner, J.M., 1975, Ventura County water resources management study aquifer delineation in the Oxnard-Calleguas area, Ventura County: Ventura County Department of Public Works Flood Control District, Technical Information Record, January 1975.

United Water Conservation District, 2003, Coastal Saline Intrusion: Report to United Water Conservation District, 25 p.

United Water Conservation District, 2008, Urban water management plan for the Oxnard-Hueneme system, Report to United Water Conservation District, 55 p.

U.S. Census Bureau, 1990, U.S. Census ftp site, accessed September 28, 2010, at ftp://ftp2.census.gov/census_1990.

U.S. Environmental Protection Agency, 1998, Code of Federal Regulations, title 40—protection of environment, chapter 1—environmental protection agency, subchapter E—pesticide programs, part 159—statements of policies and interpretations, subpart D—reporting requirements for risk/benefit information—40 CFR 159.184: Washington, National Archives and Records Administration (September 19, 1997; amended June 19, 1998), Federal Register, v. 62, no. 182, accessed September 5, 2008, at http://www.epa.gov/EPA-PEST/1997/September/Day-19/p24937.htm.

U.S. Environmental Protection Agency, 2006, 2006 Edition of the Drinking Water Standards and Health Advisories (updated August 2006): Washington, U.S. Environmental Protection Agency, Office of Water EPA/822/R-06-013, accessed September 27, 2010, at http://www.epa.gov/waterscience/criteria/drinking/dwstandards.pdf.

U.S. Environmental Protection Agency, 2008a, Drinking water contaminants: U.S. Environmental Protections Agency website, accessed June 1, 2008, at http://www.epa.gov/safewater/contaminants/index.html.

U.S. Environmental Protection Agency, 2008b, Drinking water standards and health advisories—Drinking water standards and health advisory tables, Archived drinking water standards 2006 (pdf document), accessed June 1, 2008, at http://www.epa.gov/waterscience/criteria/drinking/.

U.S. Geological Survey, 2010, What is the Priority Basin Project?: U.S. Geological Survey website, accessed September 9, 2010, at http://ca.water.usgs.gov/gama.

Vedder, J.G., Wagner, H.C., and Schoellhamer, J.E., 1969, Geologic framework of the Santa Barbara Channel Region, *in* Geology, Petroleum Development, and Seismicity of the Santa Barbara Channel Region, California: U.S. Geological Survey Professional Paper 679-A, 11 p.

Vogel, J.C., and Ehhalt, D., 1963, The use of the carbon isotopes in groundwater studies, *in* Radioisotopes in Hydrology: Tokyo, IAEA, p. 383–395.

Weber, F.H., Cleveland, G.B., Kahle, J.F., Kiessling, E.F., Miller, R.V., Mills, M.F., Morton, D.M., and Cilweck, B.A., 1973, Geology and mineral resources study of southern Ventura County, California: California Division of Mines and Geology Preliminary Report 14, 102 p.

Weber, F.H., Kiessling, E.W., Sprotte, E.C., Johnson, J.A., Sherburne, R.W., and Cleveland, G.B., 1976, Seismic hazards study of Ventura County, California: Sacramento, Calif., California Department of Conservation, California Division of Mines and Geology Open-File Report 76-5, 396 p., pls. 3A and 3B.

Williams, L.B., and Hervig, R.L., 2004, Boron isotope composition of coals—A potential tracer of organic contaminated fluids: Applied Geochemistry, v. 19, p. 1625–1636.

Wright, M.T., and Belitz, Kenneth, 2010, Factors controlling the regional distribution of vanadium in groundwater: Ground Water, v. 48, no. 4, p. 515–525.

Yeats, R.S., Huftile, G.J., and Grigsby, F.B., 1988, Oak Ridge fault, Ventura fold belt, and the Sisar decollement, Ventura basin, California: Geology, v. 16, p. 1112–1116.

Yerkes, R.F., Sarna-Wojcicki, A.M., and Lajoie, K.R., 1987, Geology and Quaternary deformation of the Ventura area, *in* Recent reverse faulting in the Transverse Ranges, California: U.S. Geological Survey Professional Paper 1339, p. 169–178.

Zogorski, J.S., Carter, J.M., Ivahnenko, T., Lapham, W.W., Moran, M.J., Rowe, B.L., Squillace, P.J., and Toccalino, P.L., 2006. Volatile organic compounds in the Nation's ground water and drinking-water supply wells: U.S. Geological Survey Circular 1292, 101 p.

Appendix A. Selection of California Department of Public Health (CDPH)-Grid Wells

California requires samples to be collected regularly from public-supply wells under Title 22 (California Department of Health Services, 2007). Historical data derived from these samples are available from the California Department of Public Health (CDPH) database. Assembly Bill (AB) 599 directs the Groundwater Ambient Monitoring and Assessment (GAMA) Program to use available data and to collect new data as needed. The GAMA Priority Basin Project uses this monitoring data along with newly collected data to characterize the water quality of the primary aquifers. The CDPH database provided additional water-quality data for the grid-based and spatially weighted approaches to estimating aquifer-scale proportions for a wide range of constituents. CDPH data were not used to provide data for grid-wells for VOCs, pesticides, or perchlorate because these constituents were sampled for by the USGS at all grid wells, and because reporting levels for these constituents in the CDPH database generally were not sufficiently low enough to differentiate between "low" and "moderate" relative-concentrations.

Of the 48 grid cells in the Santa Clara River Valley study unit (SCRV), 42 cells had USGS-grid data for organic and special interest constituents; 16 of these 42 cells also had USGS-grid data for inorganic constituents; and 6 of the 48 grid cells did not have USGS-grid data because no well was sampled (figs. A1*A* and A1*B*). Three approaches were used to select CDPH inorganic constituent data for each grid cell where the USGS did not sample for inorganic constituents.

The first approach was to select CDPH data for the grid well sampled by the USGS for other constituents, provided the CDPH data met quality-control criteria. Cation-anion balance was used as the quality-control assessment metric. Because water is electrically neutral and must have a balance between positive (cations) and negative (anions) electrically charged dissolved species, the cation/anion imbalance commonly is used as a quality-assurance check for water sample analysis (Hem, 1985). Cation-anion balance was calculated as the difference between the total cations and total anions divided by the average, expressed as a percentage:

$$percent\ difference = \left(\frac{\left| \sum cations - \sum anions \right|}{\sum cations + \sum anions} \right) * 100$$

where

$\sum$cations is the sum of calcium, magnesium, sodium, and potassium in milliequivalents per liter (meq/L) and,

$\sum$anions is the sum of chloride, sulfate, fluoride, nitrate, and bicarbonate in milliequivalents per liter (meq/L).

An imbalance, or percentage difference, greater than or equal to 10 percent can indicate uncertainty in the quality of the data. The most recent CDPH data for the USGS-grid wells with missing data were evaluated to determine whether cation-anion imbalances for the CDPH data were less than 10 percent. If so, the CDPH data for those wells were selected for use as the grid-well data for inorganic constituents. It was assumed that if analyses met acceptable-quality-control criteria for major ion data, then the data quality for the analyses at these wells also would be acceptable for trace elements, nutrients, and radiochemical constituents. This approach resulted in the selection of inorganic data from CDPH at 14 USGS-grid wells. For identification purposes, data from the CDPH for these grid wells were assigned GAMA identifications numbers equivalent to the GAMA USGS-grid well numbers, but with "DG" inserted between the study area prefix and sequence number (for example, CDPH-grid well SCRV-DG-05 is the same well as USGS-grid well SCRV-05, table A1).

If the first approach did not yield CDPH inorganic data for a grid cell, then the second approach was to search the CDPH database to identify the well with the highest rank, from the random ranking of CDPH wells during the original selection of USGS-grid wells for each cell, with a cation-anion imbalance less than 10 percent. This approach resulted in selecting CDPH inorganic data for wells not sampled by USGS for 15 grid cells. Each of these 15 CDPH-grid wells was located in the same cell as the corresponding USGS-grid well but not necessarily right next to that USGS-grid well. To identify these new CDPH-grid wells, well IDs were created that added "DPH" after the study unit prefix and then added the grid-cell number for the study unit (for example, CDPH-grid well SCRV-DPH-26).

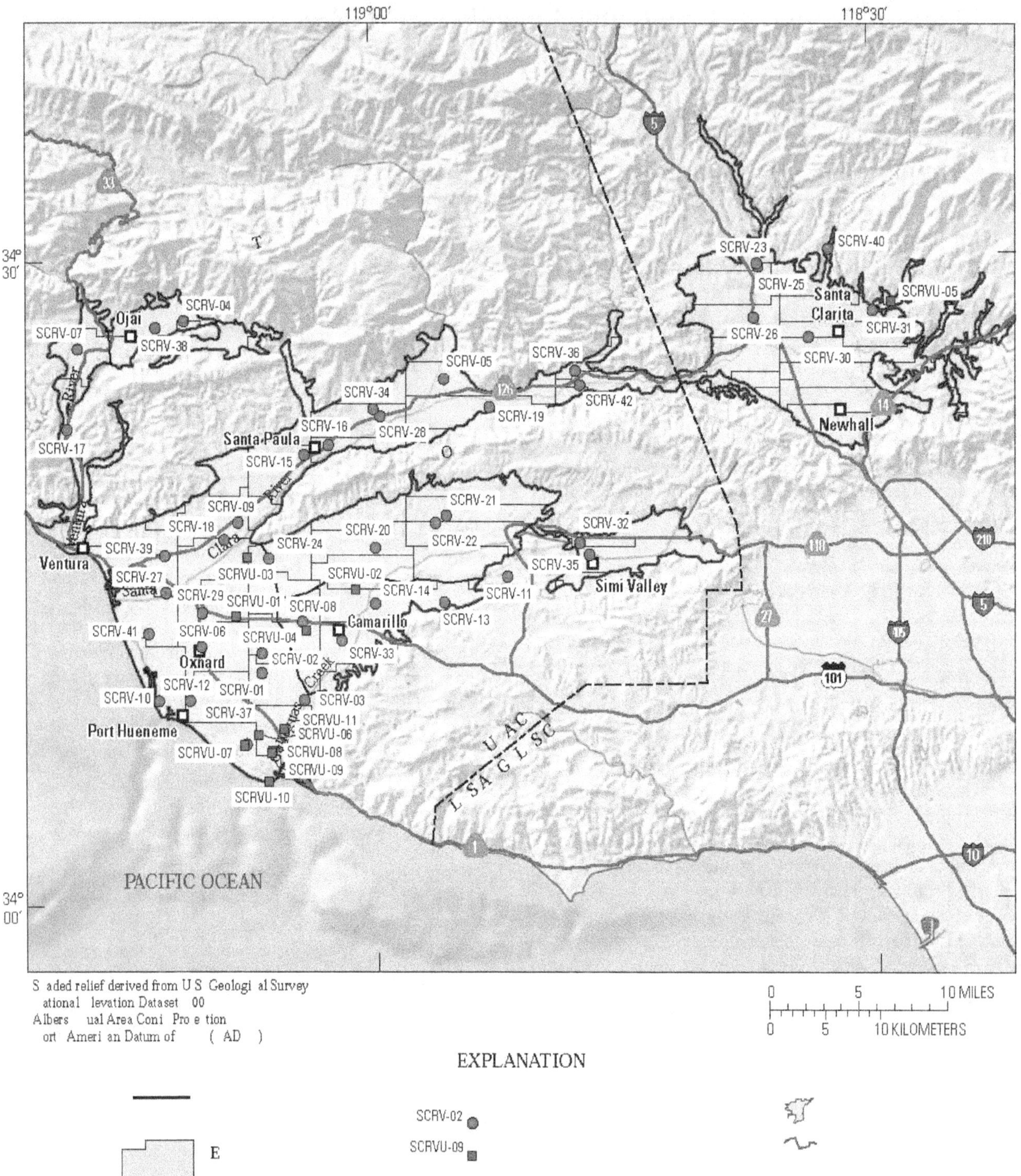

Figure A1. Identifiers and locations of (*A*) USGS-grid and USGS-understanding wells sampled during April through June 2007, and (*B*) CDPH-grid wells at which data for inorganic constituents from the California Department of Public Health were used, Santa Clara River Valley study unit, California GAMA Priority Basin Project.

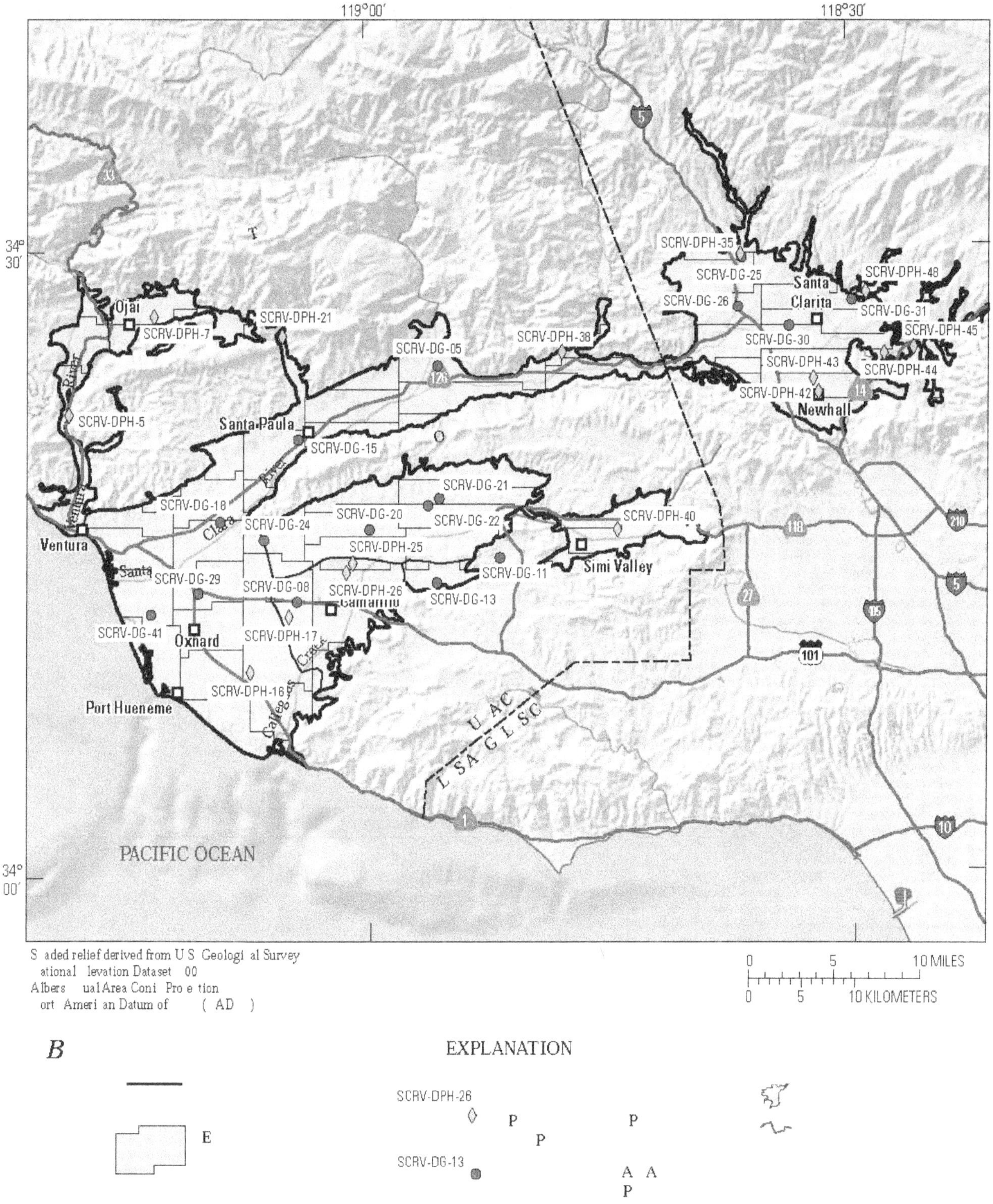

Figure A1—Continued

Table A1. Nomenclature for wells sampled by USGS or selected from the CDPH database in the Santa Clara River Valley study unit, California, GAMA Priority Basin Project.

[CDPH, California Department of Public Health; GAMA, Groundwater Ambient and Monitoring Assessment Program; USGS, U.S. Geological Survey; SCRV, Santa Clara River Valley USGS-grid well; SCRVU, USGS-understanding well; SCRV-DG, CDPH-grid well with USGS and CDPH data; SCRV-DPH, CDPH-grid well with CDPH data only; –, no data]

USGS GAMA well identification No. (USGS data only)	CDPH GAMA well identification No. (USGS and CDPH data)	CDPH GAMA well identification No. (CDPH data only)	Grid cell No.
USGS- and CDPH-grid wells			
SCRV-01	–	–	16
SCRV-02	–	–	17
SCRV-03	–	–	15
SCRV-04	–	–	8
SCRV-05	SCRV-DG-05	–	34
SCRV-06	–	–	12
SCRV-07	–	–	6
SCRV-08	SCRV-DG-08	–	18
SCRV-09	–	–	9
SCRV-10	–	–	1
SCRV-11	SCRV-DG-11	–	29
SCRV-12	–	–	13
SCRV-13	SCRV-DG-13	–	28
SCRV-14	–	–	26
SCRV-15	SCRV-DG-15	–	20
SCRV-16	–	–	23
SCRV-17	–	–	5
SCRV-18	SCRV-DG-18	–	10
SCRV-19	–	–	33
SCRV-20	SCRV-DG-20	–	24
SCRV-21	SCRV-DG-21	–	31
SCRV-22	SCRV-DG-22	–	30
SCRV-23	–	–	35
SCRV-24	SCRV-DG-24	–	19
SCRV-25	SCRV-DG-25	–	36
SCRV-26	SCRV-DG-26	–	37
SCRV-27	–	–	3
SCRV-28	–	–	22
SCRV-29	SCRV-DG-29	–	11
SCRV-30	SCRV-DG-30	–	46
SCRV-31	SCRV-DG-31	–	47
SCRV-32	–	–	40
SCRV-33	–	–	27
SCRV-34	–	–	21
SCRV-35	–	–	41
USGS- and CDPH-grid wells—Continued			
SCRV-36	–	–	38
SCRV-37	–	–	14
SCRV-38	–	–	7
SCRV-39	–	–	4
SCRV-40	–	–	48
SCRV-41	SCRV-DG-41	–	2
SCRV-42	–	–	39
–	–	SCRV-DPH-5	5
–	–	SCRV-DPH-7	7
–	–	SCRV-DPH-16	16
–	–	SCRV-DPH-17	17
–	–	SCRV-DPH-21	21
–	–	SCRV-DPH-25	25
–	–	SCRV-DPH-26	26
–	–	SCRV-DPH-35	35
–	–	SCRV-DPH-38	38
–	–	SCRV-DPH-40	40
–	–	SCRV-DPH-42	42
–	–	SCRV-DPH-43	43
–	–	SCRV-DPH-44	44
–	–	SCRV-DPH-45	45
–	–	SCRV-DPH-48	48
USGS-understanding wells			
SCRVU-01	–	–	11
SCRVU-02	–	–	26
SCRVU-03	–	–	10
SCRVU-04	–	–	17
SCRVU-05	–	–	48
SCRVU-06	–	–	14
SCRVU-07	–	–	14
SCRVU-08	–	–	15
SCRVU-09	–	–	15
SCRVU-10	–	–	15
SCRVU-11	–	–	15

If the cation-anion balance for data from the well in the CDPH database in a grid cell was not less than 10 percent, then the third approach was to select the highest randomly ranked well in the CDPH database with any of the needed inorganic data. These wells may not have met the charge-balance criteria because a complete set of major-ion data was not available to calculate a cation-anion balance. This approach resulted in selection of two USGS-grid wells (SCRV-11 and SCRV-41) from which CDPH inorganic data were used, but for which those CDPH data had not passed the cation-anion balance check. Because these wells were USGS-grid wells, a well ID was created that added “DG” to the GAMA ID (for example, well SCRV-DG-41).

The result of these approaches was one well per grid cell with data from the USGS database, the CDPH database, or both databases. Inorganic data for 31 CDPH-grid wells in the CDPH database were used (fig. A1*B*). Data were not available for all inorganic constituents from all 31 CDPH-grid wells. However, data for most of the inorganic constituents were available for 47 of the 48 grid cells. Table 3 shows the number of USGS- and CDPH-grid wells with data for each inorganic constituent.

Estimates of aquifer-scale proportions for constituents made on the basis of a small number of wells have a larger error associated with the 90 percent confidence intervals (based on the Jeffreys interval for the binomial distribution, Brown and others, 2001). Analysis of the combined data sets to evaluate the occurrence of high or moderate relative-concentrations for inorganic constituents was not affected by differences in reporting levels between USGS- and CDPH-grid data because concentrations greater than one-half of water-quality benchmarks (relative-concentration greater than 0.5) generally were substantially greater than the highest reporting levels. Comparisons between USGS- and CDPH-grid data are described in appendix B.

Appendix B. Comparison of CDPH and USGS-GAMA Data

CDPH and USGS-GAMA data were compared to assess the validity of combining data from these different sources. Because laboratory reporting levels for most organic constituents and trace elements were substantially lower for USGS-GAMA data than for CDPH data (table 4), it was not possible to directly compare concentrations of many constituents in individual wells in any meaningful way. However, concentrations of major ions and nitrate, which generally are prevalent and have concentrations substantially above reporting levels, could be compared for each well using data from both sources.

Comparisons were made for wells that were analyzed by GAMA Priority Basin Project for inorganic and radiochemical constituents that also had CDPH data within the most recent 3-year interval. Major ion and nitrate data were available for thirteen wells in the USGS and CDPH databases. The small number of wells prevented a statistically robust analysis of the paired results for each individual constituent. However, the paired analyses for nine different constituents (calcium, magnesium, potassium, sodium, alkalinity, chloride, sulfate, TDS, nitrate-N) with values above the reporting levels in both databases was a large enough dataset (72 pairs) for meaningful statistical comparison.

A non-parametric signed rank test indicated no significant differences between the paired USGS-GAMA and CDPH data ($z = 0.694$, $p = 0.4873$). Although differences between the paired datasets occurred for a few wells, most sample pairs plotted close to a 1-to-1 line (fig. B1). The relative percent difference (absolute difference of the two values divided by the average of the two values, RPD) was calculated for each data pair. The median RPD was 6.7 percent; 91 percent of the RPD values were less than 20 percent. These direct comparisons indicated that these GAMA and CDPH inorganic data were not significantly different, which gave support to the use of the CDPH data for inorganic constituents to supplement the USGS-grid well data for the cells with an incomplete suite of inorganic analyses.

Piper diagrams show the relative abundance of major cations and anions (on a charge equivalent basis) as a percentage of the total ion content of the water (fig. B2). Piper diagrams often are used to define groundwater type (Hem, 1985). Combined USGS-GAMA and CDPH major-ion data for grid wells were plotted on piper diagrams (Piper, 1944) along with all CDPH major-ion data from November 1, 2003, to October 31, 2006, to determine whether the groundwater types in grid wells were similar to groundwater types historically observed in the study unit. All cation/anion data in the CDPH database with a cation/anion balance less than 10 percent were retrieved and plotted on the piper diagrams for comparison with USGS- and CDPH-grid well data.

The ranges of water types for grid wells and other wells from the CDPH database for the current period were similar (fig. B2). Most water samples from wells were classified as *mixed cation-mixed anion* type waters; no single cation accounted for more than 60 percent of the total cations, and no single anion accounted for more than 60 percent of the total anions. The most common cation was calcium, and the major anions were sulfate and bicarbonate. Waters in a minority of wells were classified as *mixed cation-bicarbonate*, *calcium-mixed anion*, or *sodium-mixed anion* type waters, indicating that bicarbonate, calcium, or sodium accounted for more than 60 percent of the total anions and cations, respectively.

The determination that the range of relative abundances of major cations and anions in grid wells (47 wells) was similar to the range of those in the selected CDPH-other wells (151 wells) indicates that the grid wells represent most of the diversity of water types present in the SCRV. However, three minor differences between USGS- and CDPH-grid data and CDPH-other data were evident.

First, chloride is the predominant anion in one USGS-grid well (lower right of the lower right triangle of the piper diagram, fig. B2). There are no CDPH wells that plot in this area of the piper diagram. This USGS-grid well is an irrigation well near the southern tip of the SCRV where there are no active CDPH wells.

Second, sodium is the predominant cation for three CDPH-other wells (lower right of the lower left triangle of the piper diagram, fig. B2). Two of these CDPH-other wells are in the Oxnard subbasin; there were no USGS- or CDPH-grid wells that had sodium as the predominant cation. One CDPH-other well is in the Santa Clara River Valley East subbasin, where USGS-grid wells were not sampled for major ions.

Third, calcium in one CDPH-other well was greater than 75 percent (lower left of the lower left triangle of the piper diagram, fig. B2), which is much higher than in any grid well. This well is in the Santa Clara River Valley East subbasin between wells SCRV-30 and SCRV-31, both with much lower concentrations of calcium.

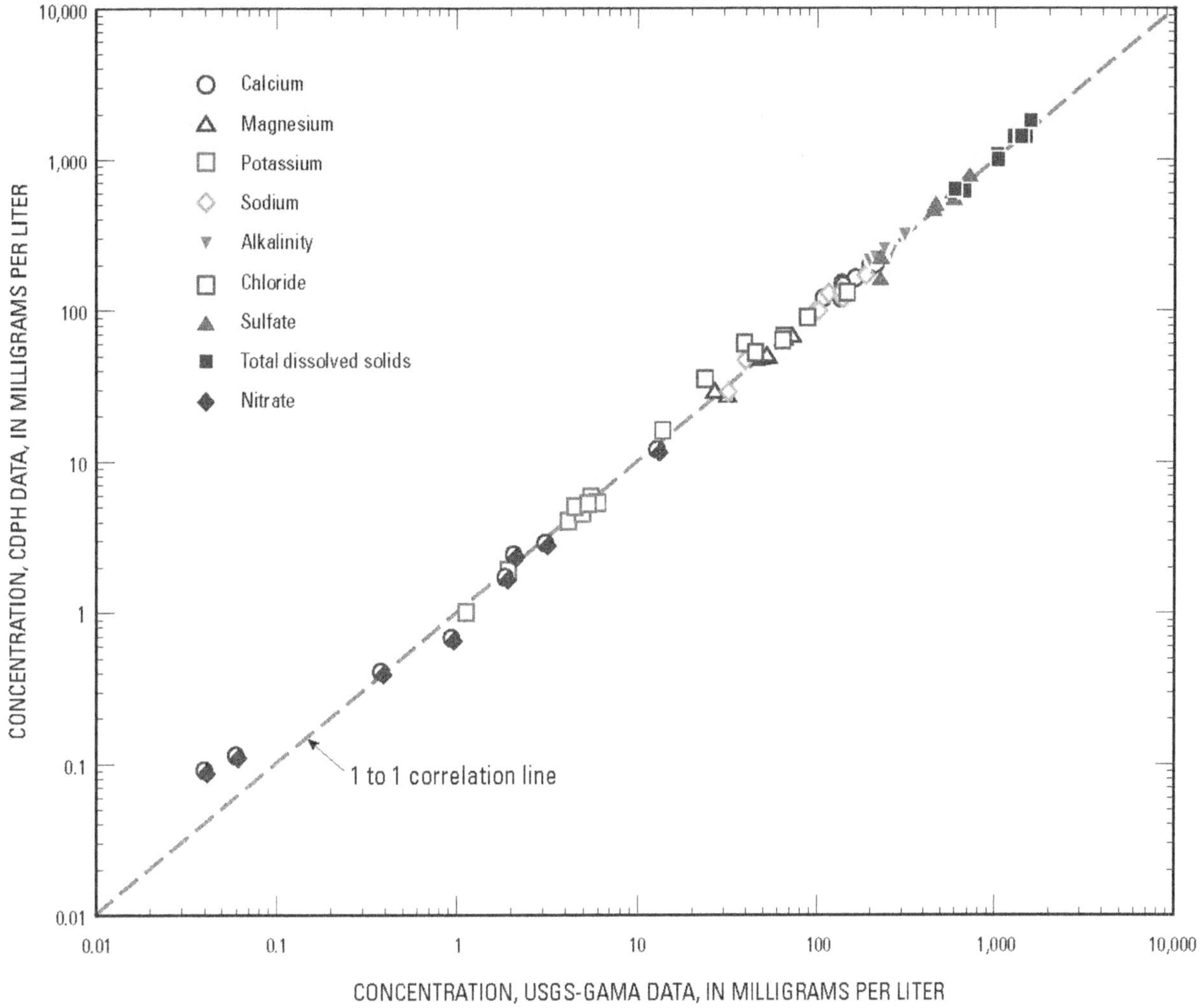

Figure B1. Paired inorganic concentrations from wells sampled by the USGS, April–June 2007, and the most recent available analysis in the California Department of Public Health for the same wells, November 1, 2003, to October 31, 2006, Santa Clara River Valley study unit, California GAMA Priority Basin Project.

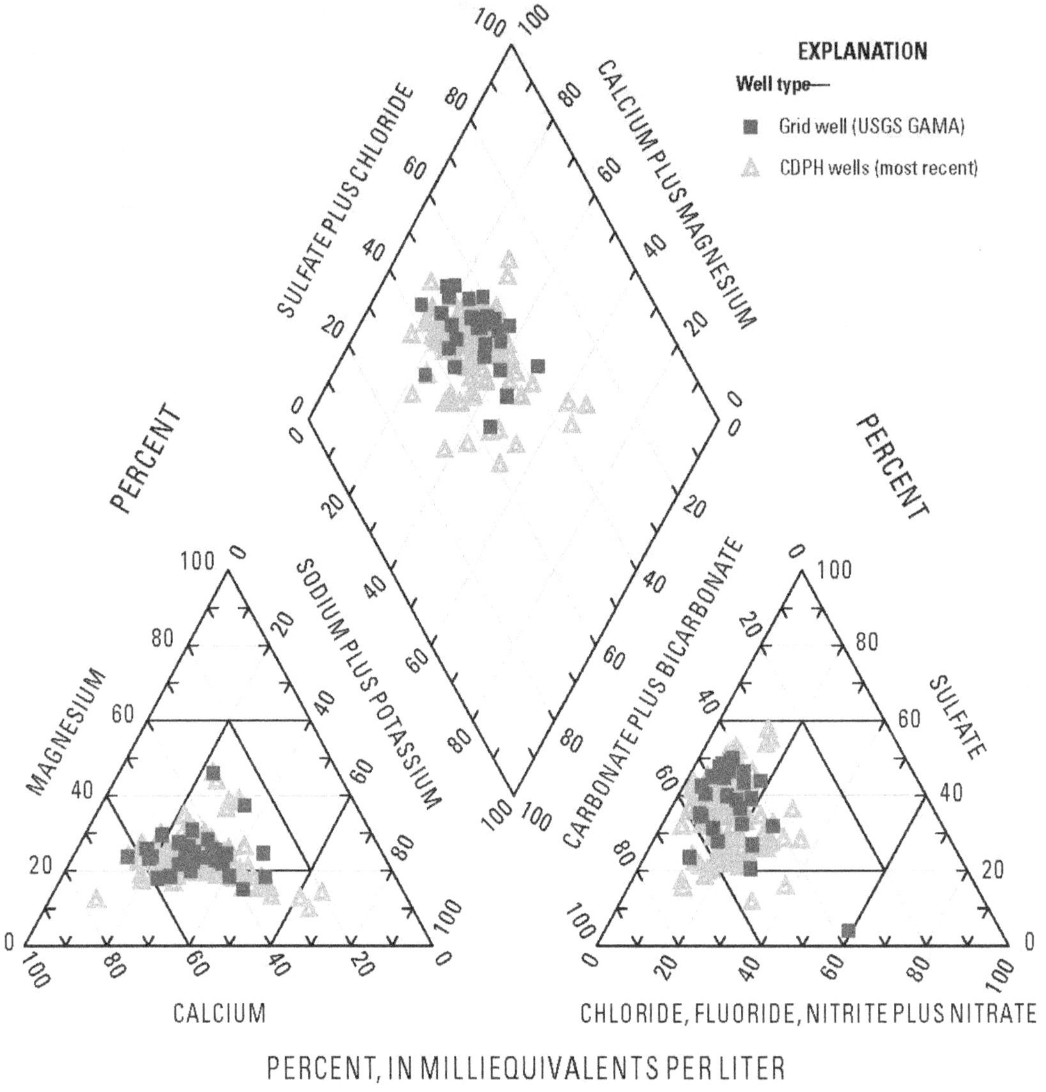

Figure B2. Comparing water types in for USGS- and CDPH-grid wells with water types in all wells in the California Department of Public Health (CDPH) database that have a charge imbalance of less than 10 percent, Santa Clara River Valley study unit, California GAMA Priority Basin Project.

Appendix C. Calculation of Aquifer-Scale Proportions

Two statistical approaches, grid-based and spatially weighted, were selected to evaluate the aquifer-scale proportions of the primary aquifers in the SCRV study unit that had high, moderate, or low relative-concentrations (concentration relative to its water-quality benchmark) of constituents. Raw detection frequencies also were calculated for individual constituents, but were not used for estimating aquifer-scale proportions because this method creates spatial bias towards regions with large numbers of wells.

Grid-Based Calculation

One well in each grid cell, a "grid well," is used to represent the primary aquifer system (Belitz and others, 2010). The relative-concentration for each constituent (concentration relative to its water-quality benchmark) was then evaluated for each grid well. The proportion of the primary aquifers with high relative-concentrations was calculated by dividing the number of cells with concentrations greater than the benchmark (relative-concentration greater than 1) by the total number of grid wells in the SCRV. Proportions containing moderate and low relative-concentrations were calculated similarly. Confidence intervals for grid-based aquifer-scale proportions were computed using the Jeffreys interval for the binomial distribution (Brown and others, 2001). The grid-based estimate is spatially unbiased. However, the grid-based approach may not identify constituents that are present at high relative-concentrations in small proportions of the primary aquifer system.

The grid-based aquifer-scale proportions for constituent classes also were calculated on a one-value–per-grid-cell basis. A cell with a high relative-concentration for any constituent in the class was defined as a high cell, and the high proportion was calculated as the number of high cells divided by the number of cells with data for any of the constituents in that class. The moderate proportion for the constituent class was calculated similarly, except that a cell already defined as high could not also be defined as moderate. A cell with a moderate relative-concentration for any constituent in the class without a high value for any constituent in the class was defined as moderate. The grid-based aquifer-scale proportion for the low category was calculated similarly, such that a cell could only be low if the relative-concentration was neither moderate nor high for any constituent in the class. The proportions for the high, moderate, and low categories were expected to total 100 percent, except for small differences as a result of rounding.

Spatially Weighted Calculation

The spatially weighted calculations of aquifer-scale proportions uses the most recent value for a constituent from all wells in the CDPH database with data in the 3-year interval prior to USGS-GAMA sampling (November 1, 2003, to October 31, 2006) in the SCRV, from all USGS-grid well data, and from selected USGS-understanding well data. The spatially weighted approach computes the aquifer-scale proportion using the percentage of wells with high relative-concentrations from all wells in each cell, instead of using data from only one well. For each constituent, the high aquifer-scale proportion was computed by calculating the proportion of wells with high relative-concentrations in each grid cell and dividing by the number of cells (Belitz and others, 2010):

$$P_i = \frac{W_{high}}{W_{total}}$$

$$P_{SU} = \frac{\sum_{i=1}^{n} P_i}{n},$$

where

P_i is the proportion of wells in the ith cell with high relative-concentrations,

W_{total} is the number of wells in the ith cell with data for the constituent,

W_{high} is the number of wells in the ith cell with high relative-concentrations,

P_{SU} is the aquifer-scale proportion for the study unit, and

n is the number of cells with data for the constituent.

Similar procedures were used to calculate the proportions of moderate and low relative-concentrations. The resulting proportions are expected to be spatially unbiased (Isaaks and Srivastava, 1989).

Raw Detection Frequencies

The raw detection frequencies of wells with high relative-concentrations for constituents were calculated using the same data used for the spatially weighted approach. This approach is the percentage (frequency) of wells in the study unit with high relative-concentrations. However, raw detection frequencies are not spatially unbiased because the wells in the CDPH database and USGS-understanding wells are not uniformly distributed. Consequently, high relative-concentrations for wells clustered in a particular area represent a small part of the primary aquifer system and could be given a disproportionately high weight compared to that given by spatially unbiased approaches. Raw detection frequencies of high relative-concentrations are provided for reference in this report but were not used to assess aquifer-scale proportions.

Appendix D. Attribution of Potential Explanatory Factors

Land-Use Classification

Land use was classified using an enhanced version of the satellite-derived (30-m pixel resolution) nationwide USGS National Land Cover Dataset (Nakagaki and others, 2007). This dataset has been used in previous national and regional studies relating land use to water quality (Gilliom and others, 2006; Zogorski and others, 2006). The dataset characterizes land cover during the early 1990s. The imagery is classified into 25 land-cover classifications (Nakagaki and Wolock, 2005). These 25 land-cover classifications were assigned to 3 general land-use classifications: urban, agricultural, and natural. Land-use statistics were assigned using USGS National Land Cover Dataset for the study unit, and for areas within a radius of 500 m around each grid, understanding, and all CDPH wells (table D1) (Johnson and Belitz, 2009). Any overlapping of the areas within a radius of 500 m around adjacent wells was not accounted for in calculating land-use statistics.

Septic Systems

Septic tank density was determined from housing characteristics data from the 1990 U.S. Census (U.S. Census Bureau, 1990, accessed September 28, 2010, at ftp://ftp2.census.gov/census_1990). The density of septic tanks in each housing census block was calculated from the number of tanks and block area. The density of septic tanks around each well was then calculated from the area-weighted mean of the block densities for blocks within 500 m around each well location. (Tyler Johnson, U.S. Geological Survey, written commun., 2009) (table D1).

Well-Construction Information

Well-construction data were from driller's logs. Other sources of well-construction data were ancillary records of well owners and the USGS National Water Information System database. Well identification verification procedures are described by Montrella and Belitz (2009). Well depths and depths to the top- and bottom-of-perforations for USGS-grid, USGS-understanding, and CDPH-grid wells are listed in table D2. Wells were classified as production, monitoring, or domestic wells (table D2). Production wells pump the groundwater from the aquifer to a distribution system. Monitoring wells tend to be short-screened wells installed exclusively for monitoring purposes. Domestic wells pump groundwater from the aquifer for home use.

Groundwater Age Classification

Groundwater dating techniques indicate the time that has elapsed after the groundwater was last in contact with the atmosphere. Techniques used to estimate groundwater residence times or 'age' include those based on tritium (^{3}H; Tolstikhin and Kamenskiy, 1969; Torgersen and others, 1979) and ^{3}H in combination with its decay product helium-3 (^{3}He) (Schlosser and others, 1988), carbon-14 (^{14}C) activities (Vogel and Ehhalt, 1963; Plummer and others, 1993: Kalin, 2000), and dissolved noble gases, particularly helium-4 accumulation (Davis and DeWiest, 1966; Andrews and Lee, 1979; Cey and others, 2008; Kulongoski and others, 2008).

Tritium (^{3}H) is a short-lived radioactive isotope of hydrogen with a half-life of 12.32 years (Lucas and Unterweger, 2000). Tritium is produced naturally in the atmosphere from the interaction of cosmogenic radiation with nitrogen (Craig and Lal, 1961), by above-ground nuclear explosions, and by the operation of nuclear reactors. Above-ground nuclear-bomb testing between 1951 and 1980 (peak production in 1963) introduced much larger quantities of ^{3}H than natural production into the atmosphere (Michel, 1989; Solomon and Cook, 2000). Tritium enters the hydrological cycle as precipitation following oxidation to tritiated water. Consequently, the presence of ^{3}H in groundwater may be used to identify water that has exchanged with the atmosphere in the past 50 years. By determining the ratio of ^{3}H to its decay product ^{3}He, the time that the water has resided in the aquifer can be calculated more precisely than by using ^{3}H alone (Takaoka and Mizutani, 1987; Poreda and others, 1988). Tritium activities and tritium-helium ages of the water samples are given in table D3.

^{14}C is a radioactive isotope of carbon with a half-life of 5,730 years that is formed naturally in the atmosphere by the interaction of cosmic-ray neutrons with nitrogen, and to a lesser degree, interaction with oxygen and carbon. ^{14}C is incorporated into carbon dioxide, which is mixed throughout the atmosphere; the carbon dioxide is dissolved in precipitation and incorporated into the hydrologic cycle. ^{14}C activity in groundwater, expressed as percent modern carbon (pmc), reflects exposure to the atmospheric ^{14}C source. ^{14}C can be used to estimate groundwater ages ranging from 1,000 to less than 30,000 years before present (Clark and Fritz, 1997). Calculated ^{14}C ages (table D3) in this study are referred to as "uncorrected" because they have not been adjusted to consider exchanges with sedimentary sources of carbon (Fontes and Garnier, 1979; Kalin, 2000). The ^{14}C age (residence time) is calculated based on the decrease in ^{14}C activity as a result of radioactive decay after groundwater recharge, relative to an assumed initial ^{14}C concentration (Clark and Fritz, 1997). A mean initial ^{14}C activity of 99 pmc was assumed for this study, with estimated errors on calculated groundwater ages as much as ±20 percent.

Table D1. Land-use classification and septic tank density for USGS-grid and -understanding wells sampled by USGS for April–June 2007, and for CDPH-grid wells for inorganic constituents, Santa Clara River Valley study unit, California, GAMA Priority Basin Project.

[Land-use classification is based on a 500-meter radius around each well (Johnson and Belitz, 2009). Density of septic tanks in a circle with a 500-meter radius around each well (U.S. Census Bureau, 1990). CDPH, California Department of Public Health; USGS, U.S. Geological Survey; SCRV, Santa Clara River Valley USGS-grid well; SCRVU, USGS-understanding well; SCRV-DG, CDPH-grid well with USGS and CDPH data; SCRV-DPH, CDPH-grid well with CDPH data only; km, kilometer, –, no data]

USGS GAMA well identification No.	CDPH GAMA well identification No.	Land use (percent)			Land-use classification	Septic density (tanks/km^2)
		Agricultural	Natural	Urban		
USGS- and CDPH-grid wells						
SCRV-01	–	84.8	7.3	7.9	Agricultural	3.8
SCRV-02	–	88.2	6.2	5.6	Agricultural	1.6
SCRV-03	–	88.4	9.6	1.9	Agricultural	1.5
SCRV-04	–	68.0	2.4	29.6	Agricultural	52
SCRV-05	SCRV-DG-05	19.7	48.8	31.5	Mixed	0.3
SCRV-06	–	0.0	6.3	93.7	Urban	16
SCRV-07	–	31.0	26.9	42.0	Mixed	22
SCRV-08	SCRV-DG-08	67.4	17.0	15.7	Agricultural	27
SCRV-09	–	81.4	14.7	3.9	Agricultural	2.5
SCRV-10	–	0.0	20.8	79.2	Urban	0.2
SCRV-11	SCRV-DG-11	59.0	41.0	0.0	Agricultural	15
SCRV-12	–	0.2	2.1	97.7	Urban	0.0
SCRV-13	SCRV-DG-13	75.4	21.9	2.7	Agricultural	12
SCRV-14	–	37.8	10.7	51.5	Urban	4.3
SCRV-15	SCRV-DG-15	0.1	5.5	94.4	Urban	6.7
SCRV-16	–	8.2	42.0	49.7	Mixed	4.8
SCRV-17	–	0.2	81.6	18.2	Natural	13
SCRV-18	SCRV-DG-18	53.6	8.1	38.3	Agricultural	33
SCRV-19	–	49.0	50.5	0.5	Mixed	1.4
SCRV-20	SCRV-DG-20	87.1	12.5	0.5	Agricultural	8.2
SCRV-21	SCRV-DG-21	84.5	15.5	0.0	Agricultural	8.2
SCRV-22	SCRV-DG-22	86.3	13.4	0.3	Agricultural	8.2
SCRV-23	–	17.2	55.2	27.6	Natural	0.3
SCRV-24	SCRV-DG-24	75.3	16.5	8.2	Agricultural	3.6
SCRV-25	SCRV-DG-25	26.3	55.7	18.0	Natural	0.6
SCRV-26	SCRV-DG-26	26.2	41.8	32.0	Mixed	8.0
SCRV-27	–	57.0	33.4	9.5	Agricultural	0.0
SCRV-28	–	71.6	22.0	6.4	Agricultural	6.8
SCRV-29	SCRV-DG-29	2.7	12.0	85.2	Urban	137
SCRV-30	SCRV-DG-30	14.4	79.2	6.4	Natural	0.6
SCRV-31	SCRV-DG-31	0.7	35.6	63.7	Urban	6.7
SCRV-32	–	0.0	50.5	49.5	Mixed	0.1
SCRV-33	–	70.7	5.6	23.7	Agricultural	1.5
SCRV-34	–	44.8	54.2	1.0	Natural	6.9
SCRV-35	–	0.0	8.7	91.3	Urban	23
USGS- and CDPH-grid wells—Continued						
SCRV-36	–	36.0	14.1	49.9	Mixed	5.7
SCRV-37	–	57.2	33.0	9.9	Agricultural	2.4
SCRV-38	–	55.2	11.6	33.2	Agricultural	18
SCRV-39	–	15.9	11.5	72.6	Urban	11
SCRV-40	–	1.5	98.5	0.0	Natural	0.4
SCRV-41	SCRV-DG-41	72.4	24.2	3.4	Agricultural	1.3
SCRV-42	–	65.2	34.6	0.2	Agricultural	4.3
–	SCRV-DPH-5	0.0	52.6	47.4	Natural	65
–	SCRV-DPH-7	55.2	11.6	33.2	Agricultural	18
–	SCRV-DPH-16	87.6	12.4	0.0	Agricultural	4.0
–	SCRV-DPH-17	33.0	25.0	42.0	Mixed	0.7
–	SCRV-DPH-21	0.0	100.0	0.0	Natural	6.3
–	SCRV-DPH-25	70.9	28.9	0.2	Agricultural	73
–	SCRV-DPH-26	12.0	7.6	80.4	Urban	59
–	SCRV-DPH-35	21.1	47.0	32.0	Mixed	0.4
–	SCRV-DPH-38	40.4	40.2	19.4	Mixed	5.0
–	SCRV-DPH-40	0.0	2.3	97.7	Urban	33
–	SCRV-DPH-42	0.3	4.1	95.5	Urban	52
–	SCRV-DPH-43	7.0	53.4	39.6	Natural	45
–	SCRV-DPH-44	0.1	37.2	62.7	Urban	16
–	SCRV-DPH-45	0.8	80.3	18.9	Natural	11
–	SCRV-DPH-48	15.1	46.8	38.0	Mixed	5.3
USGS-Understanding Wells						
SCRVU-01	–	45.4	5.8	48.8	Mixed	17
SCRVU-02	–	43.3	9.6	47.1	Mixed	65
SCRVU-03	–	80.9	11.0	8.1	Agricultural	5.0
SCRVU-04	–	49.9	29.2	20.8	Mixed	1.2
SCRVU-05	–	33.4	40.9	25.7	Mixed	3.8
SCRVU-06	–	78.4	19.2	2.4	Agricultural	2.3
SCRVU-07	–	54.2	37.1	8.7	Agricultural	2.3
SCRVU-08	–	0.1	57.3	42.6	Natural	5.1
SCRVU-09	–	0.1	57.3	42.6	Natural	5.1
SCRVU-10	–	0.0	86.6	13.4	Natural	3.2
SCRVU-11	–	93.9	5.6	0.5	Agricultural	1.4

Table D2. Well type and construction information for USGS-grid and -understanding wells sampled April–June 2007, and CDPH-grid wells for inorganic constituents, Santa Clara River Valley study unit, California, GAMA Priority Basin Project.

[Well depth determined from completed well depth reported on driller's log. Top of perforations determined from depth to bottom of solid (unperforated) casing reported on driller's log. CDPH, California Department of Public Health; USGS, U.S. Geological Survey; SCRV, Santa Clara River Valley USGS-grid well; SCRVU, USGS-understanding well; SCRV-DG, CDPH-grid well with USGS and supplemental CDPH data; SCRV-DPH, CDPH-grid well with CDPH data only; ft, foot; LSD, land-surface datum; na, not available]

USGS GAMA well identification No.	CDPH GAMA well identification No.	Well type	Construction information		
			Well depth (ft below LSD)	Top-of-perforations (ft below LSD)	Bottom-of-perforations (ft below LSD)
USGS- and CDPH-grid wells					
SCRV-01	–	Production	1,300	590	1,280
SCRV-02	–	Production	1,310	750	1,290
SCRV-03	–	Production	1,023	443	1,003
SCRV-04	–	Production	na	na	na
SCRV-05	SCRV-DG-05	Production	300	50	280
SCRV-06	–	Production	220	120	220
SCRV-07	–	Production	242	92	232
SCRV-08	SCRV-DG-08	Production	910	700	890
SCRV-09	–	Production	863	703	863
SCRV-10	–	Production	766	610	738
SCRV-11	SCRV-DG-11	Production	636	316	636
SCRV-12	–	Production	1,200	720	1,180
SCRV-13	SCRV-DG-13	Production	399	204	375
SCRV-14	–	Production	830	512	740
SCRV-15	SCRV-DG-15	Production	670	452	653
SCRV-16	–	Production	700	260	700
SCRV-17	–	Production	60	15	60
SCRV-18	SCRV-DG-18	Production	420	300	400
SCRV-19	–	Production	100	na	na
SCRV-20	SCRV-DG-20	Production	1,440	800	1,440
SCRV-21	SCRV-DG-21	Production	980	650	na
SCRV-22	SCRV-DG-22	Production	980	670	980
SCRV-23	–	Production	na	na	na
SCRV-24	SCRV-DG-24	Production	1,042	642	1,042
SCRV-25	SCRV-DG-25	Production	203	60	165
SCRV-26	SCRV-DG-26	Production	142	na	na
SCRV-27	–	Production	873	403	853
SCRV-28	–	Production	334	105	240
SCRV-29	SCRV-DG-29	Production	252	107	na
SCRV-30	SCRV-DG-30	Production	208	na	na
SCRV-31	SCRV-DG-31	Production	150	56	150
SCRV-32	–	Production	300	50	230
SCRV-33	–	Production	504	224	504
SCRV-34	–	Production	920	500	920
SCRV-35	–	Production	300	70	290
SCRV-36	–	Production	300	na	na
SCRV-37	–	Production	541	421	521

Table D2. Well type and construction information for wells sampled by USGS for April–June 2007, and CDPH-grid wells for inorganic constituents, Santa Clara River Valley Groundwater Ambient Monitoring and Assessment (GAMA) study unit, California.—Continued

[Well depth determined from completed well depth reported on driller's log. Top of perforations determined from depth to bottom of solid (unperforated) casing reported on driller's log. CDPH, California Department of Public Health; USGS, U.S. Geological Survey; SCRV, Santa Clara River Valley USGS-grid well; SCRVU, USGS-understanding well; SCRV-DG, CDPH-grid well with USGS and supplemental CDPH data; SCRV-DPH, CDPH-grid well with CDPH data only; ft, foot; m, meter; LSD, land-surface datum; na, not available]

USGS GAMA well identification No.	CDPH GAMA well identification No.	Well type	Construction information		
			Well depth (ft below LSD)	Top-of-perforations (ft below LSD)	Bottom-of-perforations (ft below LSD)
USGS- and CDPH-grid wells—Continued					
SCRV-38	–	Production	644	281	644
SCRV-39	–	Production	1,190	580	1,080
SCRV-40	–	Domestic	na	na	na
SCRV-41	SCRV-DG-41	Production	300	na	na
SCRV-42	–	Production	275	110	275
–	SCRV-DPH-5	Production	29	20	29
–	SCRV-DPH-7	Production	644	281	644
–	SCRV-DPH-16	Production	200	150	190
–	SCRV-DPH-17	Production	1,071	801	1,051
–	SCRV-DPH-21	Production	na	na	na
–	SCRV-DPH-25	Production	830	467	830
–	SCRV-DPH-26	Production	884	564	864
–	SCRV-DPH-35	Production	na	na	na
–	SCRV-DPH-38	Production	300	na	na
–	SCRV-DPH-40	Production	600	350	600
–	SCRV-DPH-43	Production	1,290	485	1,280
–	SCRV-DPH-42	Production	na	na	na
–	SCRV-DPH-44	Production	na	na	na
–	SCRV-DPH-45	Production	na	na	na
–	SCRV-DPH-48	Production	302	102	302
USGS-understanding wells					
SCRVU-01	–	Production	820	400	820
SCRVU-02	–	Production	759	459	759
SCRVU-03	–	Production	330	100	320
SCRVU-04	–	Production	1,483	403	1,463
SCRVU-05	–	Production	na	na	na
SCRVU-06	–	Monitor	740	720.0	740
SCRVU-07	–	Monitor	720	680	720
SCRVU-08	–	Monitor	640	600	640
SCRVU-09	–	Monitor	970	930	970
SCRVU-10	–	Monitor	220	200	220
SCRVU-11	–	Production	810	400	810

Table D3. Groundwater age classification information for wells sampled by the USGS for April–June 2007, Santa Clara River Valley study unit, California, GAMA Priority Basin Project.

[Samples classified as pre-modern if recharged prior to 1955. Samples classified as modern if recharged after 1955. Samples classified as mixed if sample contains both modern and pre-modern water. GAMA, Groundwater Ambient Montoring and Assessment Program; SCRV, Santa Clara River Valley USGS-grid well; SCRVU, USGS-understanding well; >, greater than; <, less than]

USGS GAMA well identification No.	Tritium activity (TU)	Tritium-helium age (years)	Uncorrected carbon-14 age (years)	Percentage of terrigenic helium in total helium	Groundwater age classification
USGS-grid wells					
SCRV-01	0.00	>50		24.9	Pre-modern
SCRV-02	0.19	>50		70.2	Pre-modern
SCRV-03	0.0	>50	16,400	74.8	Pre-modern
SCRV-04	2.10	10.1		0.0	Modern
SCRV-05	1.82	3.8		0.0	Modern
SCRV-06	2.10	11.9		0.0	Modern
SCRV-07	1.60	Not datable	<1,000	0.0	Mixed
SCRV-08	0.00	>50		53.5	Pre-modern
SCRV-09	0.09	>50		0.0	Mixed
SCRV-10	0.09	>50	3,800	0.0	Mixed
SCRV-11	0.50	>50		1.8	Mixed
SCRV-12	0.31	>50	7,600	12.5	Pre-modern
SCRV-13	2.60	16.2		11.9	Mixed
SCRV-14	0.00	>50		47.2	Pre-modern
SCRV-15	1.00	19.2		0.0	Modern
SCRV-16	1.60	8.2		2.2	Modern
SCRV-17	2.10	Not datable		0.0	Modern
SCRV-18	0.09	>50		6.5	Pre-modern
SCRV-19	2.51	9.7		0.0	Modern
SCRV-20	0.00	>50		0.0	Mixed
SCRV-21	0.19	>50		10.3	Pre-modern
SCRV-22	0.09	>50		15.2	Pre-modern
SCRV-23	2.82	5.1		0.0	Modern
SCRV-24	2.10	8.1		0.0	Modern
SCRV-25	2.41	Not datable		2.5	Modern
SCRV-26	3.10	>50		19.5	Mixed
SCRV-27	1.00	23.6		0.0	Modern
SCRV-28	1.10	17.7		0.0	Modern
SCRV-29	3.61	3.8		0.0	Modern
SCRV-30	2.82	>50		9.5	Mixed
SCRV-31	3.20	>50		25.1	Mixed
SCRV-32	1.00	22.3		33.1	Mixed
SCRV-33	1.41	Not datable		4.6	Modern
SCRV-34	0.00	>50		0.0	Mixed
SCRV-35	2.82	Not datable	3,300	3.5	Mixed
SCRV-36	2.60	Not datable		0.0	Modern

Table D3. Groundwater age classification information for wells sampled by the USGS for April–June 2007, Santa Clara River Valley study unit, California, GAMA Priority Basin Project.—Continued

[Samples classified as pre-modern if recharged prior to 1955. Samples classified as modern if recharged after 1955. Samples classified as mixed if sample contains both modern and pre-modern water. GAMA, Groundwater Ambient Montoring and Assessment program; SCRV, Santa Clara River Valley USGS-grid well; SCRVU, USGS-understanding well; >, greater than; <, less than]

USGS GAMA well identification No.	Tritium activity (TU)	Tritium-helium age (years)	Uncorrected carbon-14 age (years)	Percentage of terrigenic helium in total helium	Groundwater age classification
USGS-grid wells—Continued					
SCRV-38	0.60	>50		0.0	Mixed
SCRV-39	0.00	>50		3.7	Mixed
SCRV-41	2.32	16.1		0.0	Modern
SCRV-42	3.61	3.0		4.9	Modern
USGS-understanding wells					
SCRVU-01	2.01	25.1	1,300	0.0	Mixed
SCRVU-02	2.70	>50	3,100	66.5	Mixed
SCRVU-03	1.97	1.3	<1,000	0.0	Mixed
SCRVU-04	0.19	>50	11,300	86.2	Pre-modern
SCRVU-05	3.01	>50		16.7	Mixed
SCRVU-06	0.00	>50		75.0	Pre-modern
SCRVU-10	0.41	>50		64.4	Pre-modern
SCRVU-11	0.00	>50		93.7	Pre-modern

Helium (He) is a naturally occurring inert gas produced by the radioactive decay of lithium, thorium, and uranium in the earth. Measured He concentrations in groundwater are determined as the sum of air-equilibrated He, He from dissolved-air bubbles, terrigenic He, and tritiogenic ^3He. The helium (^3He and ^4He isotopes) concentrations in groundwater commonly exceed the expected solubility equilibrium values, which are a function of the temperature of the water, as a result of subsurface production of both isotopes, and their subsequent release into the groundwater (Morrison and Pine, 1955; Andrews and Lee, 1979; Torgersen, 1980; Andrews, 1985; Torgersen and Clark, 1985). The presence of terrigenic He in groundwater, from its production in aquifer material or deeper in the crust, is indicative of long groundwater residence times. The amount of terrigenic He is defined as the concentration of the total measured He, minus He from air equilibration and dissolved-air bubbles. Percentage of terrigenic He is defined as the concentration of terrigenic He (as defined previously) divided by the total measured He in the sample (corrected for air-bubble entrainment). Samples in which more than 5 percent of the total He is terrigenic He (percentage of terrigenic He) indicate groundwater has a residence time of more than 100 yrs.

Recharge temperatures for 48 samples were determined from dissolved neon, argon, krypton, and xenon data using methods described by Aeschbach-Hertig and others (1999). The only modeled recharge temperatures accepted were those for which the probability was greater than 1 percent, and the sum of the squared deviations between the modeled and the measured concentrations (weighted with the experimental 1-sigma errors) was equal to or greater than the observed value (Aeschbach-Hertig and others, 2000). The recharge temperature with the highest probability for each sample was used in this report. For samples that had a modeled recharge temperature probability of less than 1 percent, the mean recharge temperature of all samples was used.

The groundwater age was computed using $^3H/^3He$ as described by Poreda and others (1988). The $^3He/^4He$ of samples was determined by the linear regression of the percentage of terrigenic He and δ^3He ($[\delta^3He = Rmeas/Ratm - 1] \times 100$ percent) of samples with less than 1 tritium unit. Rmeas is the ratio of $^3He/^4He$ in the measured sample; Ratm is the ratio of $^3He/^4He$ in the atmosphere. Calculations of the recharge temperature using noble gases and $^3He/^4He$ are useful because they can be used to constrain helium-based groundwater ages further.

In this study, the ages of samples were classified as pre-modern, modern, and mixed (table D3). Groundwater with ^{3}H activity less than 1 tritium unit (TU), percentage of terrigenic He greater than 5 percent, and ^{14}C less than 90 pmc was designated as pre-modern. Pre-modern groundwater was defined as having been recharged before 1950. Groundwater with ^{3}H greater than 1 TU, percentage of terrigenic He less than 5 percent, and ^{14}C greater than 90 pmc was designated as modern. Modern groundwater is defined as having been recharged after 1950. Samples with pre-modern and modern components are designated as mixed groundwater. In reality, pre-modern groundwater could contain small fractions of modern groundwater and modern groundwater could contain small fractions of pre-modern groundwater. Previous investigations have used a range of tritium values from 0.3 to 1.0 TU as thresholds for distinguishing pre-1950 from post-1950 groundwater (Michel, 1989; Plummer and others, 1993, p. 260; Michel and Schroeder, 1994; Clark and Fritz, 1997, p. 185; Manning and others, 2005). By using a tritium value of 1.0 TU for the threshold in this study, the age classification scheme allows a larger fraction of modern groundwater to be classified as pre-modern than if a lower threshold was used. A lower threshold for tritium would result in fewer samples classified as pre-modern age than as mixed age, when other tracers, ^{14}C and terrigenic He, indicated that the samples primarily were pre-modern age. This higher threshold was considered more appropriate for this study because many of the wells were production wells with long screens and mixing water of pre-modern and modern age likely occur.

Geochemical Conditions

Geochemical conditions investigated as potential explanatory variables in this report include oxidation-reduction characteristics, dissolved-oxygen concentrations, and pH (table D4). Oxidation-reduction (redox) conditions influence the mobility of many organic and inorganic constituents (McMahon and Chapelle, 2008). Along groundwater flow paths, redox conditions commonly proceed along a well-documented sequence of terminal electron acceptor processes (TEAP); one TEAP typically is predominant at a particular time and aquifer location (Chapelle and others, 1995; Chapelle, 2001). The predominant TEAPs are oxygen-reduction (oxic), nitrate-reduction, manganese-reduction, iron-reduction, sulfate-reduction, and methanogenesis. The presence of redox-sensitive chemical species indicating more than one TEAP may indicate (1) the discharge from the well includes mixed waters from different redox zones upgradient of the well, (2) the well is screened across more than one redox zone, or (3) there is spatial heterogeneity in microbial activity in the aquifer. In addition, different redox couples often are not consistent, indicating electrochemical disequilibrium in groundwater (Lindburg and Runnels, 1984) complicating the assessments of redox conditions.

In this report, redox conditions were represented in two ways: as dissolved oxygen concentration and as redox category based on the predominant TEAP(s). Dissolved-oxygen concentrations were measured at USGS-grid and USGS-understanding wells (Montrella and Belitz, 2009), but are not reported in the CDPH database (table D4). Redox conditions were classified based on dissolved oxygen, nitrate, manganese, iron, and sulfate concentrations using the classification scheme of McMahon and Chapelle (2008) (table D4). An automated workbook program was used to assign the redox class to each sample (Jurgens and others, 2009). For wells without USGS inorganic constituent data, the most recent data within the previous 3 years (November 1, 2003, to October 31, 2006) for that well in the CDPH database were used.

Table D4. Oxidation-reduction classification and pH for wells sampled by the USGS for April–June 2007, and CDPH-grid wells for inorganic constituents, Santa Clara River Valley study unit, California, GAMA Priority Basin Project.

[Redox category and redox process determined using the algorithm of McMahon and Chapelle (2008) implemented by Jurgens and others (2009) except for samples with incomplete redox data, which were excluded from the analysis. CDPH, California Department of Public Health; USGS, U.S. Geological Survey; SCRV, Santa Clara River Valley USGS-grid well; SCRVU, USGS-understanding well; SCRV-DG, CDPH-grid well with USGS and supplemental CDPH data; SCRV-DPH, CDPH-grid well with CDPH data only; redox, oxidation-reduction; mg/L, milligrams per liter; oxic, dissolved oxygen greater than 0.5; anoxic, dissolved oxygen less than 0.5; O_2, oxygen; NO_3, nitrate reducing; Mn, manganese reducing; Fe/SO_4, iron and (or) sulfate reducing; >, greater than; nc, not collected; na, not able to determine; –, no well]

USGS GAMA well identification No.	CDPH GAMA well identification No.	pH	Oxidizing and reducing constituents					Redox category	Redox process
			Dissolved oxygen (mg/L)	Nitrate plus nitrite (mg/L)	Manganese (µg/L)	Iron (µg/L)	Sulfate (mg/L)		
USGS- and CDPH-grid wells									
SCRV-01	–	7.7	<0.2	nc	nc	nc	nc	O_2 < 0.5 mg/L	unknown
SCRV-02	–	7.6	<0.2	nc	nc	nc	nc	O_2 < 0.5 mg/L	unknown
SCRV-03	–	7.7	0.4	<0.06	66.5	76	137	anoxic	Mn
SCRV-04	–	7.0	5.0	11.2	0.4	9	225	oxic	O_2
SCRV-05	SCRV-DG-05	7.0	4.1	1.56	30.0	430	557	Mixed (oxic-anoxic)	O_2-Fe/SO_4
SCRV-06	–	7.2	0.3	12.9	17.5	14	673	anoxic	NO_3
SCRV-07	–	6.8	5.9	0.94	<0.2	<6	227	oxic	O_2
SCRV-08	SCRV-DG-08	7.5	<0.2	<0.11	54.0	260	210	anoxic	Fe/SO_4
SCRV-09	–	7.1	<0.2	<0.06	230	1,420	811	anoxic	Fe/SO_4
SCRV-10	–	7.6	<0.2	<0.06	174	906	343	anoxic	Fe/SO_4
SCRV-11	SCRV-DG-11	7.5	<0.2	nc	nc	nc	nc	O_2 < 0.5 mg/L	unknown
SCRV-12	–	7.6	0.3	0.04	188	319	367	anoxic	Fe/SO_4
SCRV-13	SCRV-DG-13	7.4	2.7	26.9	<10	<50	213	oxic	O_2
SCRV-14	–	7.4	1.8	nc	nc	nc	nc	O_2 ≥ 0.5 mg/L	unknown
SCRV-15	SCRV-DG-15	7.3	0.4	5.85	120	<50	454	anoxic	NO_3/Mn
SCRV-16	–	7.5	<0.2	1.86	396	3	448	anoxic	NO_3/Mn
SCRV-17	–	7.1	2.6	nc	nc	nc	nc	O_2 ≥ 0.5 mg/L	unknown
SCRV-18	SCRV-DG-18	7.2	0.7	<0.41	160	<100	486	Mixed (oxic-anoxic)	O_2-Mn
SCRV-19	–	7.3	7.3	2.71	0.1	3	387	oxic	O_2
SCRV-20	SCRV-DG-20	7.8	0.8	0.03	100	160	102	Mixed (oxic-anoxic)	O_2-Fe/SO_4
SCRV-21	SCRV-DG-21	7.6	0.5	0.03	64.0	110	134	Mixed (oxic-anoxic)	O_2-Fe/SO_4
SCRV-22	SCRV-DG-22	7.5	0.5	nc	nc	nc	nc	O_2 ≥ 0.5 mg/L	unknown
SCRV-23	–	7.6	<0.2	nc	nc	nc	nc	O_2 < 0.5 mg/L	unknown
SCRV-24	SCRV-DG-24	7.4	1.1	6.60	80.0	740	545	Mixed (oxic-anoxic)	O_2-Fe/SO_4
SCRV-25	SCRV-DG-25	7.4	3.0	1.11	<20	<100	174	oxic	O_2
SCRV-26	SCRV-DG-26	7.5	2.2	2.98	<20	<100	237	oxic	O_2
SCRV-27	–	7.4	0.3	3.09	243	44	572	Mixed (oxic-anoxic)	O_2-Mn
SCRV-28	–	7.1	3.2	22.2	1.9	30	483	oxic	O_2
SCRV-29	SCRV-DG-29	7.3	3.4	2.94	1.6	<40	440	oxic	O_2
SCRV-30	SCRV-DG-30	7.4	3.0	3.37	<20	<100	188	oxic	O_2
SCRV-31	SCRV-DG-31	7.4	6.1	4.77	<20	<100	183	oxic	O_2
SCRV-32	–	7.0	<0.2	nc	nc	nc	nc	O_2 < 0.5 mg/L	unknown
SCRV-33	–	7.0	2.6	15.2	11.7	<6	905	oxic	O_2
SCRV-34	–	7.3	0.7	nc	nc	nc	nc	O_2 ≥ 0.5 mg/L	unknown
SCRV-35	–	7.0	7.7	12.4	<0.2	<6	893	oxic	O_2

Table D4. Oxidation-reduction classification and pH for wells sampled by USGS for April–June 2007, and CDPH-grid wells for inorganic constituents, Santa Clara River Valley Groundwater Ambient Monitoring and Assessment (GAMA) study unit, California.—Continued

[Redox category and redox process determined using the algorithm of McMahon and Chapelle (2008) implemented by Jurgens and others (2009) except for samples with incomplete redox data, which were excluded from the analysis. CDPH, California Department of Public Health; USGS, U.S. Geological Survey; SCRV, Santa Clara River Valley USGS-grid well; SCRVU, USGS-understanding well; SCRV-DG, CDPH-grid well with USGS and supplemental CDPH data; SCRV-DPH, CDPH-grid well with CDPH data only redox, oxidation-reduction; mg/L, milligrams per liter; oxic, dissolved oxygen greater than 0.5; anoxic, dissolved oxygen less than 0.5; O_2, oxygen; NO_3, nitrate reducing; Mn, manganese reducing; Fe/SO_4, iron and (or) sulfate reducing; >, greater than; nc, not collected; na, not able to determine; –, no well]

USGS GAMA well identification No.	CDPH GAMA well identification No.	pH	Oxidizing and reducing constituents					Redox category	Redox process
			Dissolved oxygen (mg/L)	Nitrate plus nitrite (mg/L)	Manganese (µg/L)	Iron (µg/L)	Sulfate (mg/L)		
USGS- and CDPH-grid wells—Continued									
SCRV-36	–	7.4	3	nc	nc	nc	nc	$O_2 \geq 0.5$ mg/L	unknown
SCRV-37	–	7.5	<0.2	<0.06	176	330	39.5	anoxic	Fe/SO_4
SCRV-38	–	7.2	0.2	nc	nc	nc	nc	$O_2 < 0.5$ mg/L	unknown
SCRV-39	–	7.3	<0.2	0.38	253	934	719	anoxic	Fe/SO_4
SCRV-41	SCRV-DG-41	nc	4.0	4.9	nc	nc	nc	$O_2 \geq 0.5$ mg/L	unknown
SCRV-42	–	7.6	4.6	2.10	0.5	8	268	oxic	O_2
–	SCRV-DPH-5	7.0	nc	0.18	<10	<50	210	na	na
–	SCRV-DPH-7	7.1	nc	2.55	690	<50	206	na	na
–	SCRV-DPH-16	7.4	nc	<0.09	<10	<50	396	na	na
–	SCRV-DPH-17	7.4	nc	0.03	20.0	350	191	na	na
–	SCRV-DPH-21	7.5	nc	0.47	10.0	<40	160	na	na
–	SCRV-DPH-25	7.1	nc	<0.11	240	1,100	880	na	na
–	SCRV-DPH-26	7.5	nc	<0.09	130	120	296	na	na
–	SCRV-DPH-35	7.6	nc	<0.45	<20	<100	124	na	na
–	SCRV-DPH-38	7.1	nc	2.98	10	530	450	na	na
–	SCRV-DPH-40	7.1	nc	9.71	<1	<40	620	na	na
–	SCRV-DPH-42	7.3	nc	2.98	nc	nc	133	na	na
–	SCRV-DPH-43	7.2	nc	3.5	<20	<100	137	na	na
–	SCRV-DPH-44	7.4	nc	5.74	<20	<100	113	na	na
–	SCRV-DPH-45	7.3	nc	3.16	<20	120	137	na	na
–	SCRV-DPH-48	7.8	nc	13.8	nc	nc	107	na	na
USGS-understanding wells									
SCRVU-01	–	7.0	1.1	2.07	5.5	7	461	oxic	O_2
SCRVU-02	–	7.2	<0.2	<0.06	149	137	586	anoxic	Fe/SO_4
SCRVU-03	–	7.6	7.9	2.34	0.6	13	518	oxic	O_2
SCRVU-04	–	7.6	<0.2	<0.06	27.6	128	316	anoxic	Fe/SO_4
SCRVU-05	–	7.4	5.1	nc	nc	nc	nc	$O_2 \geq 0.5$ mg/L	unknown
SCRVU-06	–	7.4	<0.2	<0.06	155	62	214	anoxic	Mn
SCRVU-07	–	7.1	<0.2	<0.06	2,370	7,960	2.25	anoxic	Fe/SO_4
SCRVU-08	–	6.9	<0.2	<0.06	186	<6	635	anoxic	Mn
SCRVU-09	–	7.3	<0.2	<0.06	221	<6	330	anoxic	Mn
SCRVU-10	–	7.2	0.4	<0.06	784	10,200	2,060	anoxic	Fe/SO_4
SCRVU-11	–	7.3	nc	<0.06	8.9	17	247	na	na

www.ingramcontent.com/pod-product-compliance
Lightning Source LLC
Chambersburg PA
CBHW081549170526
45166CB00009B/2635